PADDINGTON TO BIRKENHEAD

Dedication

James Henry Baker

PADDINGTON TO BIRKENHEAD

THE GREAT WESTERN'S NORTH WEST FRONTIER

Michael H.C. Baker

Pen & Sword
TRANSPORT
AN IMPRINT OF PEN & SWORD BOOKS LTD.
YORKSHIRE – PHILADELPHIA

First published in Great Britain in 2025 by
Pen and Sword Transport
An imprint of
Pen & Sword Books Ltd.
Yorkshire - Philadelphia

Copyright © Michael H.C. Baker, 2025

ISBN 978 1 39908 786 5

The right of Michael H.C. Baker to be identified as author of this work has been asserted by him in accordance with the Copyright, Designs and Patents Act 1988.

A CIP catalogue record for this book is available from the British Library.

All rights reserved. No part of this book may be reproduced or transmitted in any form or by any means, electronic or mechanical including photocopying, recording or by any information storage and retrieval system, without permission from the Publisher in writing.

Typeset in Palatino 11/13 by SJmagic DESIGN SERVICES, India.

Printed and bound in India by Replika Press Pvt. Ltd.

Pen & Sword Books Ltd. incorporates the imprints of Pen & Sword Books: After the Battle, Archaeology, Atlas, Aviation, Battleground, Discovery, Family History, History, Maritime, Military, Politics, Select, Transport, True Crime, Fiction, Frontline Books, Leo Cooper, Praetorian Press, Seaforth Publishing, Wharncliffe and White Owl.

For a complete list of Pen & Sword titles please contact

PEN & SWORD BOOKS LIMITED
George House, Beevor Street, Off Pontefract Road, Hoyle Mill, Barnsley,
South Yorkshire, England, S71 1HN.
E-mail: enquiries@pen-and-sword.co.uk
Website: www.pen-and-sword.co.uk

or

PEN AND SWORD BOOKS
1950 Lawrence Rd, Havertown, PA 19083, USA
E-mail: uspen-and-sword@casematepublishers.com
Website: www.penandswordbooks.com

CONTENTS

Acknowledgements 6

Introduction 7

Chapter 1   The Rivals 17

Chapter 2   Euston to Oxford 24

Chapter 3   Oxford to Banbury 31

Chapter 4   Banbury to Birmingham 41

Chapter 5   Birmingham 52

Chapter 6   West Midlands 62

Chapter 7   Shropshire Branch Lines 70

Chapter 8   Wolverhampton to Shrewsbury 76

Chapter 9   Shrewsbury 84

Chapter 10   The Shrewsbury to Crewe Line 97

Chapter 11   Towards Chester 114

Chapter 12   Destination 126

Chapter 13   Birkenhead Explored 134

ACKNOWLEDGEMENTS

I would like to thank many people who have contributed to this book, even though they may not at the time have realised it! Sadly and inevitably many are not with us any longer. Firstly, all my relations, not least my mother, who had the good sense to be born and live in Shropshire. Then there is John Villers, with whom I worked so closely for many years on many projects, mostly concerned with things Western, and will for ever be missed. Brian Searle is a friend who shared many outings in the Super Saloons when part of the Great Western Society Vintage Train. The contribution which that Society has made to survival of all things Great Western is incalculable, but in that context Frank Dumbleton is the name which springs first to mind. The original four 16-year-old school boy founders would have been five if the award winning Frank had not been away at school and any visit to the home of the Great Western Society Didcot seems to have something missing if he and his trusty camera (he is a superb photographer) are not present.

Peter Waller is another with whom I have worked for decades, whilst the unique Ray Stenning of Best Impressions has always been ready with advice and encouragement and despite having taught him as a sixth former very, many, many years ago, I still find his approach to design and layout incomparable.

INTRODUCTION

It is 210½ miles from Paddington to Birkenhead. It is 193½ miles from Euston to Liverpool. And something less than a mile across the Mersey from Liverpool to Birkenhead. The London and North Western Railway (LNWR) and its London, Midland and Scottish (LMS) successor employed its finest express locomotives on the London to Liverpool route, as did the Great Western Railway (GWR) between London and Birkenhead. One might reasonably suppose that there was great rivalry between the two for the traffic over these parallel lines. But that is only until one consults the timetables and finds that in steam days, the fastest Euston to Liverpool train took 3 hours 45 minutes, whilst that of its rival – if they really were rivals – took 5 hours 35 minutes. Clearly there could have been no serious competition when the Euston route had a 1 hour 50 minute advantage over the Paddington one. Trains left Paddington for Birkenhead at ten past the hour, five times a day. There was double that number, generally speaking, between Euston and Liverpool for which there was clearly a demand. One might then ask, why did the GWR and its Western Region successor bother to compete at all? The answer is that it did not pretend to be a serious rival over the whole route, although it certainly called one of its trains 'The Belfast Boat Express', suggesting that it offered a connection with the ferry which sailed for Ireland from Liverpool. Rather, its chief business was with passengers arriving at or departing from places en route, notably Banbury, Leamington Spa, Birmingham, Wolverhampton, Shrewsbury and Chester.

Birmingham, England's second city, was certainly worth fighting over and equality was pretty well achieved when in 1906 the GWR opened its cut off route by way of High Wycombe and Princes Risborough, bypassing the slower. original Oxford route. The fastest train on either route now took two hours. Thus it remained, until electrification came to the West Coast main line, including Birmingham and Wolverhampton, in the 1960s. For a brief period, whilst electrification works were inevitably causing delays, much of the traffic between London and Birmingham was concentrated on the line out of Paddington. Following the inauguration of electrification of the Euston route, there was a drastic pruning of that to and from Paddington. The last through train between Birkenhead and Paddington ran on 4 March 1967 and Birmingham Snow Hill closed in 1972. The London terminus for journeys to and from Shrewsbury became Euston with a change of trains at the former LNWR/LMS High Level Wolverhampton station.

Closing Snow Hill was a dreadful mistake. Common sense gradually re-emerged, concerns for the environment became predominant, and a smaller, and barely adequate – but better than nothing – four platform Snow Hill reopened in October 1987. Moor Street, once restricted to the suburban network south of Birmingham and also slated for closure, instead had two extra through platforms provided and a wonderfully imaginative makeover which received several awards.

Over the southern part of the old GWR northern main line, the establishment of the ambitious and go-ahead Chiltern Railways at the end of the 1990s brought about a renaissance. In May 1993, the company began to operate between Birmingham Moor Street and Marylebone, the latter a terminus which too had very nearly been closed before its strategic value and position, let alone its attractive ambiance, had been realised. Trains ran every hour, with more during peak periods. In September 1995 the line north from Snow Hill reopened to Kidderminster and Worcester.

The GWR had never really embraced electrification. In the London suburban area, the Underground's Central tube line extension, planned before 1939, took over the stopping services as far as West Ruislip in 1948, but that was all. Decades later, electrification resurfaced. In October 1993, Merseyrail began running electric trains over the final eleven mile section of the original Paddington to Birkenhead line, that between Chester and Rock Ferry, connecting there with the long established tunnel beneath Birkenhead centre and the River Mersey and so into the heart of Liverpool. The headway was a mere fifteen minutes, a huge improvement, on what had sufficed before. In 1999, trams replaced trains on the old GWR main line between Birmingham Snow Hill and Wolverhampton.

Before Cardiff officially became the Welsh capital, a ballot was held in 1955 and Shrewsbury, on account of providing the best rail link between north and south Wales, received two votes. It is true that part of the Paddington to Birkenhead line actually passes through Wales, between Chirk and Rossett, but the fact that today Shrewsbury station proclaims Arriva Wales to be its owner cannot be used as evidence that the wishes of some to become the capital of the Principality have been fulfilled.

The Paddington to Birkenhead route was always one of great contrasts, steeped in industry between Birmingham and Wolverhampton and on the approach to Birkenhead, serving the upmarket stockbroker belt between Denham and High Wycombe, and around Solihull, passing through fields where the ancient Medieval strip pattern can still be detected around Banbury, connecting ancient centres of civilisation such as Oxford, Shrewsbury and Chester, deeply rural through the Chilterns and Shropshire, within sight of mountains on the approach to Wales, and twice crosses from one country to another: not bad.

Paddington platform 1, c1935. Streamlining is all the rage, hence the bulbous nose applied to No.6014 *King Henry VII*. A picture which suggests that it isn't just boys who have found the steam locomotive an object of fascination. (Mullinger)

INTRODUCTION • 9

Edge Hill at the opening of the Liverpool and Manchester Railway. It was the very first all steam, passenger and goods carrying railway in the world. (Author's collection)

Sansaw Hall is the home of the Bibby ship owning family of Liverpool. The Shrewsbury to Crewe line runs through immediately behind the house. (Author)

Stainer 5MT 4-6-0 No. 45189 has just passed Yorton with a Liverpool to West of England express on 13 September 1961. Beyond the trees is Sansaw Hall. (Author)

The Praed Street entrance to Paddington., 2015. (Author)

2-6-2T No.6111 departs Radley in the Summer of 1956 with an Oxford bound stopping train. (Author)

An I.K. Brunel lookalike, explaining the finer points of his replica GWR broad gauge locomotive *Iron Duke*. The original was built in 1847. (Author)

A somewhat bedraggled party trying to make the best of things on the wet Paddington Arm of the Grand Union Canal. (Author)

An Arriva Wales train from South Wales pulls into Shrewsbury station, over the River Severn in the spring of 2022. Beyond is the Severn Bridge signal box, a grade one listed building, the largest mechanical signal box in the world. (Author)

Birkenhead Woodside, impressive but gloomy, c1950. (Author's collection)

Sometime in the 1930s, two LNWR 0-8-0s meet, hauling long lines of wooden, open coal wagons on the West Coast main line south of Rugby. (Author's collection)

A section of the only known surviving broad gauge carriage, dating from 1849, in the Bristol Industrial Museum. (Author)

Two preserved LNWR locomotives, No.1054, a Webb 0-6-2T, and No. 49395 a G2 0-8-0, on the Pontypool & Blaenavon Railway in 2002. (Author)

The final run of the Great Western Society's Vintage Train of ten GWR design carriages, hauled by Castle class No.5051 *Earl Bathurst* approaching Stratford-upon-Avon on 6 January 1980. (Whetmath)

No.46243, minus *City of Lancaster* nameplates, awaits breaking up at the Wigan Wagon Works in 1965. (Author)

Chapter 1
THE RIVALS

The empire occupied by the Great Western Railway was neatly encompassed within the West of England main line from Paddington to Penzance and its north western main line from Paddington to Birkenhead. But whilst a measurable number of passengers would regularly make the entire journey by the former route, practically no-one would use the latter from end to end, it being very much quicker to take a LNWR express from Euston to Liverpool Lime Street and thence a ferry from the Pier Head, or a Mersey Railway electric train from Central underground to their destination. The GWR used to advertise the 4.10pm out of Paddington as the 'Belfast Boat Express' and passengers would reach Liverpool Landing Stage via Birkenhead Woodside at 11.23pm. The 'Merseyside Express' left Euston at 6.00pm and reached Liverpool Lime Street at 10.01pm so it was clearly no contest. Where the GWR scored was in picking up Birkenhead-bound passengers along the line from Birmingham or Wolverhampton onwards. Rather ironically, the Irish ferry to Dublin today leaves from the Birkenhead bank of the Mersey, not Liverpool.

Like nearly all our main lines that ran between Paddington and Birkenhead it was made up of a number of individual, sometimes separately owned, lines which gradually amalgamated until they became one. Out of Paddington, both GWR routes shared the same tracks as far as Didcot. Here they diverged, that to Bristol and beyond continuing due west, the Birkenhead one turning north-west, first to the ancient university city of Oxford, and then on through the West Midlands, eventually reaching what was by the mid-nineteenth century, the UK's second city, Birmingham, having overtaken Bristol, which can reasonably claim to be the birthplace of the GWR. Naturally enough, such a possible honeypot attracted the attention of numerous railway entrepreneurs, the eventual outcome being that three of the largest of all the UK railway companies, the GWR, the Midland, and the LNWR, all established themselves in Birmingham. The Midland Railway also reached Bristol, something the LNWR never managed, at least not directly. We have already met the LNWR and it will never be far away; it did, after all, call itself the Premier Line, but then the GWR knew it was God's Wonderful Railway. Best let Solomon decide the outright winner.

It was possible to make money by investing in railways, although it was almost as easy to lose it. Might as well put money on a horse in the Grand National as to expect to come up with a precise estimate for building a railway; nothing much changes; think HS2. It wasn't just unqualified chancers who got it wrong. Robert Stephenson thought that he could build the line between London and Liverpool for £21,000 per mile; in the event he was out by more than double that much; Brunel made an even greater miscalculation on the London to Bristol route.

Beyond Birmingham lay what became known as the Black Country, the most heavily industrialised of any part of GWR territory. Wolverhampton marked the northern extremity of this; beyond lay deeply rural Shropshire. This is certainly the perception of that county, not least on account of the writings of P.G. Wodehouse. In fact, the area southwest of Wellington had as good a claim as anywhere on the planet to be the cradle of the Industrial

Revolution. Shrewsbury, almost formed into an island by the looping River Severn, is within sight of Wales and would find itself the most important railway junction linking North and South Wales. Twenty-two miles further on the line would actually cross into Wales, and encounter more industry, specifically coal mining in the Wrexham area, before returning to England at Chester. Chester General was a joint GWR and LNWR affair, as was the final section of the line to Birkenhead.

My loyalties were equally divided. It was the GWR which would convey us from Paddington to Shrewsbury but the final few miles to Hadnall were on what had once been LNWR metals. Initially I did not know this, but having examined the works of Ian Allan, C.J. Allen, and Hamilton Ellis, I discovered that my assumption that Hadnall had once been a Midland Railway station was wrong. It was an understandable mistake for I knew it first as LMS and then London Midland Region (LMR) property and for decades, ever since the Grouping of 1923, the Midland influence had become all pervasive, first in the liveries it applied, admittedly very fine ones, to all carriages and locomotives, and then to locomotive design, a disastrous dead end from which it was only rescued by the cavalry from Swindon in the figure of W.A. Stanier. I was to learn that the line between Shrewsbury and Crewe, with which I would become so familiar, could boast LNWR origins; these I considered far more exotic than anything the Midland Railway could offer.

Not least, I suppose, because it called itself the Premier Line. So assured was it of this that it didn't even bother to put its initials on its engine's tenders, but also because, in one sense its fall had been so great that it could said to be of tragic proportions. Whilst the Edwardian vintage Churchward Stars and Saints could still be called upon to keep time with the heaviest expresses, deep into old age in early BR days, there was only one top of the range, four cylinder ex LNWR Claughton express passenger locomotive left on the outbreak of the Second World War, and this, No.6004, managed to eke out its days mainly on freight work from Edge Hill shed. The Claughtons were the archetypical curate's egg locomotives, sometimes capable of excellent work but largely regarded as not quite up to the mark.

Just as the final section of the GWR's route into London might have been shared by the Grand Junction Railway, predecessor of the LNWR, so its final dozen or so miles, from Chester to Birkenhead, was also a joint GWR/LNWR concern. Whilst the big, powerful GWR 4700 2-8-0s would work all the way from Birkenhead to Paddington with the overnight fitted freight, no regular passenger train ever did this and thus there was no guarantee if you boarded a train of GWR carriages at Birkenhead Woodside that Swindon designed motive power would be found at its head, until reversal at Chester. Birkenhead shed was home well into the mid-1950s of the last Dean-designed pannier tanks, and also LMS 3F 0-6-0Ts, there were Stanier and Fairburn 2-6-4Ts and Collett 51xx 2-6-2Ts, Halls, Granges, Crabs and Black 5s all seemingly content to share accommodation. However, towards the end of steam the LMR gained ascendancy and in the last days of the through Paddington service Stanier 5MT 4-6-0s had a virtual monopoly between Chester and Shrewsbury, certain drivers taking the opportunity of a last fling, managed to regularly average over 60mph between Gobowen and Shrewsbury. .

A George V makes a spectacular exit from Euston in the last days of the LNWR. These fast and powerful 4-4-0s were the backbone of long distance passenger traffic on the West Coast main line at the time of Grouping. (LNWR Society)

Castle and King 4-6-0s stand at Paddington arrival platforms c1952. On the left is No.4096 *Highclere Castle*, on the right No.6005 *King George II*. (Morrison)

Liverpool's magnificent waterfront in 1962. In the foreground is the *Wallasey*, the last steam powered Mersey ferry, withdrawn the following year, on the far left the Isle of Man ferry. Later declared a world heritage site, the intrusion of too many, undistinguished, high rise buildings brought about the loss of this accolade in 2022. (Author)

The elegant roof of Liverpool Lime Street station, in the spring of 2023. (Author)

A King class 4-6-0 speeds through the Chilterns north of High Wycombe in the summer of 1948 with a Paddington to Birkenhead express.
(Morrison)

A Claughton four cylinder 4-6-0 heads north with a lengthy express out of Euston, c1925. The locomotive retains its LNWR black livery whilst its train is an eclectic mix of former Midland and LNWR carriages and brand new LMS ones.
(LNWR Society)

Appropriately named Royal Scot No.46119 'The Lancashire Fusilier' has charge of the 2.35pm Euston to Liverpool Lime Street express leaving Watford Tunnel on 30 March 1959. (Author)

Late Victorian rebuilt GWR pannier tanks worked the docks at Birkenhead for over fifty years. Withdrawal had begun long ago but there was still a need for them into the mid-1950s. Here a couple await breaking up at Swindon in 1951. (Author's collection)

Electrification of the West Coast main line out of Euston began in 1960. Around this time, the new city of Milton Keynes, embracing the onetime carriage works of the LNWR at Wolverton, was created. What is now a 7-platform station serving 7million passengers annually, south of Wolverton, was opened, in February, 1982 and here, a few months later a class 82 electric locomotive is seen approaching Milton Keynes with a. southbound parcels train. (Author)

CHAPTER 2
EUSTON TO OXFORD

Chronologically we must first of all consider Euston, the terminus of the London and Birmingham Railway. This opened in 1837, the year Queen Victoria came to the throne, and was the first railway station in London north of the Thames, the very first of all being the London Bridge terminus of the London and Greenwich Railway in 1836. It was George and Robert Stephenson who chose the Euston site, initially in 1831, which was changed several times, one proposal being that it should be beside the Grand Union Canal at Islington. Inevitably, given the established importance of canals, they play a considerable part in the very earliest histories of the two London termini directly serving Birkenhead; and, indeed, right up to the present day. The first train into Paddington did not arrive until 31 May 1838 and at that stage Paddington was still in the throes of construction, alongside the Paddington arm of the Grand Union Canal. The comparison with Euston did the incomplete Paddington no favours. Although some doubted that railways had a future, this certainly did not apply to the directors of the London and Birmingham. Their attitude was that Euston station deserved a fitting gateway to the whole new world which the railway was to open up and commissioned from their architect Philip Hardwick 'a grand but simple portico, which they considered well adapted to the national character of the undertaking'. This was the famous 70ft high Euston arch. Incredibly when the old Euston station – generally no great loss – was replaced in 1962 the arch went too, despite protests led by John Betjeman and others. Such was the shock and outrage that it ensured that nothing like it should happen again. It probably saved St Pancras, and the fact that more by luck than judgement the contemporary London and Birmingham Railway Curzon Street terminus at Birmingham was still in existence, has enabled it to become the centrepiece in the plans for the HS2 station there.

Brunel would, of course, have noted the statement made by the Stephensons and the London and Birmingham directors re the grandeur of Euston station, and although he and Robert Stephenson were great friends they were also rivals, so he determined that the completed Paddington would be a station to stand the test of time. Which, unlike Euston, it certainly has. Large, covered stations usually came in one of two forms, those with pitched roofs and those with arched ones. The great advantage of the latter was that a high, arched roof allowed the smoke and steam from locomotives to disperse much more easily. Aesthetically it was also more pleasing; and in Paris it enabled artists such as Claude Monet to produce great works of art. But it was a new concept, and generally more expensive and early stations tended to be of the pitched roof variety. This applied to Euston.

The contrast between the grandeur of the entrance and the much less impressive interior was a marked feature as technology advanced. Thus, the later Paddington and St Pancras scored over the very early Euston and its later piecemeal additions. The most striking example of this contrast could be found in Dublin right into the 1970s where the Kingsbridge station, renamed Heuston in 1966, remained a virtually perfect replica of the original Euston for, while it had quite the finest façade of any station in Ireland, the interior had just two main platforms, with several sidings between them and a low pitched roof, creating a very gloomy atmosphere. Just how perfect a time warp this was is highlighted by the makers of the film, *The First Great Train Robbery* choosing it in 1978 to represent London Bridge station, which, as we have noted earlier, dated from 1836. To be fair to Irish Rail, a €170million facelift in 2002, helped by the disappearance of steam some decades earlier, transformed Heuston into a vastly more welcoming, light filled, nine platform station: although an arched roof would have been nice.

The finished Paddington station was partly the work of Brunel, the detail that of Matthew Digby Wyatt. The first train departed on 16 January 1854, from a then incomplete station, and it would be some months before it was the finished object. The great arched roof was formed of three spans, was 699 feet long and covered four platforms although there was plenty of room for more to be added over the years. It was, at the time of its opening, the largest roofed station in the world. All grand stations in those early days featured a hotel and Paddington's was the work of Philip Charles Hardwick the son of the architect of the Euston arch. Its style was described as French classical, and it possessed 104 rooms. Brunel never did things by halves.

At this stage, the GWR was a purely broad gauge railway and the speed of its locomotives and the comfort of its wide bodied carriages confirmed how superior this was over what the rest of the country could achieve on 4ft 8½in wide tracks. Sadly, the broad gauge would eventually be defeated by sheer force of the 4ft, 8½in crowding in on it but for the moment we will consider the thrust of the Great Western's broad gauge tracks to the ancient university city of Oxford and beyond. The first bill for a line off the main West of England route at Didcot was put forward in 1837, but was defeated, as was another a year later, one of the objectors being Oxford University which feared that direct access to London might result in students 'forming improper marriages and other illegitimate connexions'. Then there was the Duke of Wellington, the university's chancellor, who had the more general objection to railways that the lower orders, 'might move about more'. Thank goodness he didn't live to see the creation of the Ian Allan Locospotters club. What the objectors hadn't realised was that both the students and the lower orders had cunningly discovered that only ten miles distant was Steventon station on the GWR main line and had flocked there, and that most of the 77,500 passengers and the 12,600 tons of freight handled at that station in 1838 originated in Oxford. Admitting defeat, the opposition withdrew and the line from Didcot to Oxford opened on 12 June 1844, *Jackson's Oxford Journal* declaring that despite 'one of these rampageous, dragonading, fire-devils' being in charge 'not one hair more than was deemed proper by each spectator' was harmed.

The far-seeing directors of the London and Birmingham Railway commissioned their architect Philip Hardwick to construct an arch at the entrance to Euston station which would provide a 'Grand Avenue for travelling between Metropolis and the Midlands and Northern parts of the United Kingdom' which he did, in 1838. 116 years later British Railways ordered it to be knocked down despite a tremendous fight put up Nicholas Pevsner, John Betjeman, and others to save it. (Author's collection)

John Betjeman and the author gazing in awe at the magnificent great arch of St. Pancras station, which very nearly went the same way as the Euston arch. The combination of Sir George Gilbert Scott's 300-room railway hotel and William Henry Barlow's single span arch, then the largest in the world , provided what many consider the epitome of Victorian architectural and engineering achievement. Yet in the 1960s there were proposals to demolish both. This time, Betjeman and a growing chorus of appalled dissent prevailed. The decision to make St Pancras the London terminus of Eurostar was nothing short of inspirational. (Maeve Baker)

A northbound stopping train from Euston has just emerged from Watford Tunnel sometime in the 1930s. The locomotive is one of the appropriately named Watford Tanks, a long-lived Webb design of 1898-1902 vintage, which lasted until 1952. (LNWR Society)

The working replica of Stephenson's *Rocket* at Bold Colliery in 1980 during the 150th anniversary celebrations of the opening of the Liverpool and Manchester Railway. Beyond is a class 56 diesel locomotive attached to the Advanced Passenger Train. (Author)

Firefly, the 2005 Didcot-built replica of the first really successful GWR broad gauge locomotive. Designed by Daniel Gooch, who was appointed by Brunel at the age of 20, there were sixty-three members of the class which ran from 1840 until 1879. Gooch eventually became GWR chairman and lived until 1889. (Author)

It is not just film companies, with their vast resources, which can bring back past times. Some of the longest established, most enthusiastic and best run, amateur preservation groups have achieved wonders. One such is the Great Western Society. (Author)

The Great Western Society was founded by four 16-year-old schoolboys, in 1962. Since then, at its base at Didcot Railway Centre, it has grown beyond any of their wildest dreams. Here we see a recreated broad gauge scene, with a replica Brunel era locomotive hauling jolly passengers in period attire. (Author)

The most powerful GWR express passenger locomotives were the thirty members of the King class, dating from 1927. No.6012 *King Edward VI* has charge of the 13-coach Paddington to Shrewsbury express as it accelerates past Old Oak Common, 4 April, 1958. (Author)

A 'rampageous, dragonading, fire devil' Irish version at Downpatrick. (Author)

Chapter 3
OXFORD TO BANBURY

The joined up railway from Paddington to Birkenhead was completed on 1 October 1861. It had taken seventeen years. Nothing unusual about that. It is a story repeated again and again, so many of the great trunk routes started out as a local initiative which grew and grew, sometimes by accident, sometimes by design, gradually coming together until a complete railway all the way from London to wherever; although not always by the quickest route. Which was why, many years later, the GWR built a line through the Chilterns avoiding Oxford and also several cut offs on the main line to the West.

Oxford, home to 'the world's oldest English speaking school' was clearly somewhere which needed to be connected to the capital as soon as possible so, as the GWR extended its empire northwards as well as to the west, it was inevitable that the university city was its first objective, although even if it was not on a precise straight line between London and Birmingham. The GWR reached Oxford on 12 June 1844, opening a station at Western Road, Grandpoint. This vanished long ago to be replaced by one on the present site, a half mile distant from the city centre, and also where the GWR met the Oxford and Rugby Railway. This was promoted by the GWR as a means of reaching deep into Grand Junction territory and thus gaining access to the north of England. It was taken over by the GWR on 4 May 1846, some two months before the Grand Junction became part of the LNWR. Clearly the last thing the LNWR wanted was this incursion by its deadly rival and eventually the line, which was single track initially, settled for terminating at Banbury. Trains began running on 2 October 1850. The LNWR, in the meantime, opened its own station at Banbury, Merton Street, some four months earlier, the terminus of a line from Bletchley on its main route from Euston to the north.

I have in front of me a rather delightful, and very much of its time 1913 Adam and Charles Black publication telling the story of the creation of the GWR's main lines in its series, 'Peeps at Great Railways'. Passing Reading there is a reference to its abbey 'now, alas! little more than a very dull and black collection of shapeless walls', whilst its first abbot, Hugh, we are told 'was quartered in the disgustingly barbarous fashion of Tudor times'. With a shudder we move on hurriedly to Didcot Junction and 'the great trunk route to the North which branches off to Oxford, Warwick and Birmingham, making the name of a very humble village familiar to everyone who travels on the Great Western'.

Several pages later we are brought up to date, 1913 style, with the direct route to Birmingham, which 'first goes through some pretty unspoiled country in a part of Buckinghamshire only lately invaded by a railway'. Noting that Oxford 'is now on a connecting rather than on a main line', that High Wycombe is 'a pleasant old market-town' and that 'Bicester, a little town in famous hunting country', we regain the original Paddington to Birkenhead line at Banbury, 'the home of the cakes to be found in most bakers' shops'.

From here on all the way to Birkenhead the route is by the early Great Western. All it has to say of the 'districts of Birmingham and Wolverhampton' is that 'The country seen from the train is very largely given up to factories and works of all descriptions, and there is much needless and quite extravagant ugliness and untidiness on every side'. Untidiness, my dear – look the other way.

The track through Banbury was doubled within two years, and Banbury would very quickly become a junction of considerable importance. The line was built to the broad gauge and this extended northwards through Leamington Spa, and Birmingham to Wolverhampton, that being its northern limit. However, as the UK rail network grew ever more extensive and complex, the GWR had to face reality and allow more and more of its routes to be of mixed gauge. Indeed, an Act of Parliament was passed on 8 August 1846 prohibiting the building of any line which was not of 4ft 8½in. The writing was on the wall for the broad gauge and its fate was sealed. By 1869 it had vanished north of Oxford and would finally come to an end on the weekend of 20-22nd May 1892.

The final broad gauge train left Penzance at 9.10pm on 21 May that year, and terminated at Exeter at four next morning, conveying Inspector Scantlebury who handed the stationmaster a piece of paper which read, 'This is the last broad gauge train to travel over the line between Penzance and Exeter' and that enabled 5,000 men 'with saws, crowbars, and shovels' each in a gang of twenty to set to work. We are told that the men 'brought their own food, and the Railway Company provided great quantities of oatmeal, water and sugar, from which a thin gruel was made, and this quenches the thirst of the hard-working permanent–way men…By Saturday night the narrowing was almost complete,' and the following day the first through train from Carlisle, which had travelled by way of Crewe, Shrewsbury, Hereford and Bristol, composed of 'smaller London and North Western carriages, arrived in South Devon'. It could be argued that the LNWR had finally conquered but not really so; the result was a draw for at the same time, GWR carriages now took the same route, not only to Carlisle but also Manchester and Liverpool and, a single coach, to Glasgow. This, in 1932, left Penzance at 12.30 pm and got to its destination at 6.45 next morning.

At Banbury there was a connection between the LNWR and the GWR stations, but it was used only for transfer movements of freight wagons and never saw passenger workings. A vastly more important junction was that at Aynho, some six miles south of Banbury where the original line from Didcot and Oxford met the new Bicester/High Wycombe line in 1910, shortening the Paddington–Banbury distance from 81 miles 12 chains to 62 miles 33 chains, an important advantage in the competition with the LNWR for the lucrative Birmingham traffic. The Aynho junction remains, although both Aynho stations are long closed, Aynho for Deddington on the Oxford line on 2 November 1964 and Aynho Park on 7 November the year before. Some of the buildings of both remain.

There was another junction between Banbury and Aynho, at Kings Sutton, a station which still exists. Here a line to Kingham, which connected with the GWR's London to Worcester main line, was opened in 1887 and this connected with a line to Andoversford and Cheltenham which had been completed six years earlier. In 1906, a flying junction was constructed at Kingham which meant that through trains between Banbury and Cheltenham no longer had to reverse. In the meantime, in 1899, the Great Central Railway (GCR) had opened its main line from the Midlands to Marylebone. This made a number of connections, notably a six mile one from Culworth Junction to Banbury. The GWR and the GCR, who had developed a very cordial relationship, in 1906 inaugurated the Newcastle to Swansea and Barry Ports to Ports Express over this route. Despite taking advantage of the new flyover at Kingham, this was not speedy, for most of the Banbury to Cheltenham line was single track.

A journey over any part of it, but especially between Kingham and Cheltenham, took one through a part of England which then – and it hasn't much changed – was the epitome of all that tourist boards and prime ministers, such as John Major and David Cameron, saw as the perfect English rural scene, an immaculately laid out, gently rolling, manicured, countless shades of green, countryside, in which a blackbird sings all day, the sort where P.G. Wodehouse's Bertie Wooster would relax when not knocking off policeman's helmets in London. Kingham was one station down the line from Adlestrop about which Edward Thomas wrote, following a journey when his train stopped there, briefly on 24 June 1914, the poem which has become the best remembered and most perfect evocation of rural England immediately before the First World War.

The Culworth Junction to Banbury line proved of enormous strategic value in both world wars as a link between the Midlands, the North of England and the south coast ports. The GCR and, later the London North Eastern Railway (LNER), took turns with the GWR to provide the carriages for the Ports to Ports Express. Sadly, under nationalisation the Great Central's main line into Marylebone fell into the clutches of the LMR which, inevitably perhaps, favoured its own, and argued, with Beeching looming, that the GCR route merely duplicated what the former Midland Railway main line could easily provide and the London Extension, the Culworth to Banbury link and the Kings Sutton to Andoversford lines all closed in 1962. Marylebone itself came perilously near to closure, so near that the closure notices were posted around the station, but it survived, and as we shall see in later chapters, the twenty-first century has seen its transformation. Praise the Lord!

Oxford c1920. A Birkenhead to Bournemouth train stands in the up platform. The locomotive is 4-4-0 No. 3304 *River Tamar*, built in 1898. The three leading carriages belong to the London and South Western Railway. The middle one is a clerestory roof dining car, indicating that this was considered a train of some importance. There is still a regular service between Oxford and Bournemouth, refreshments a thing of the past. (Author's collection)

The recreated coal mine at the Black Country Museum, Dudley, with a preserved Midland Red bus in the foreground. Many of the lesser used branch lines in the vicinity of the Paddington to Birkenhead route between Banbury and Gobowen succumbed to the Midland Red between the end of the Second World War and the 1970s. Founded in 1978, the Black Country museum has assembled some fifty shops, houses, industrial buildings, depots, garages, etc from the Metropolitan Boroughs of Dudley, Sandwell, Walsall and the City of Wolverhampton. Which together form the Black Country. (Author)

Spring time in the Chilterns, March 2002. Chiltern Railways 3-car unit No.168110 speeds towards Saunderton on a working from Marylebone to Banbury. These units originated in 1997 and are the backbone of the efficient and popular Chilterns Railways fleet. (Author)

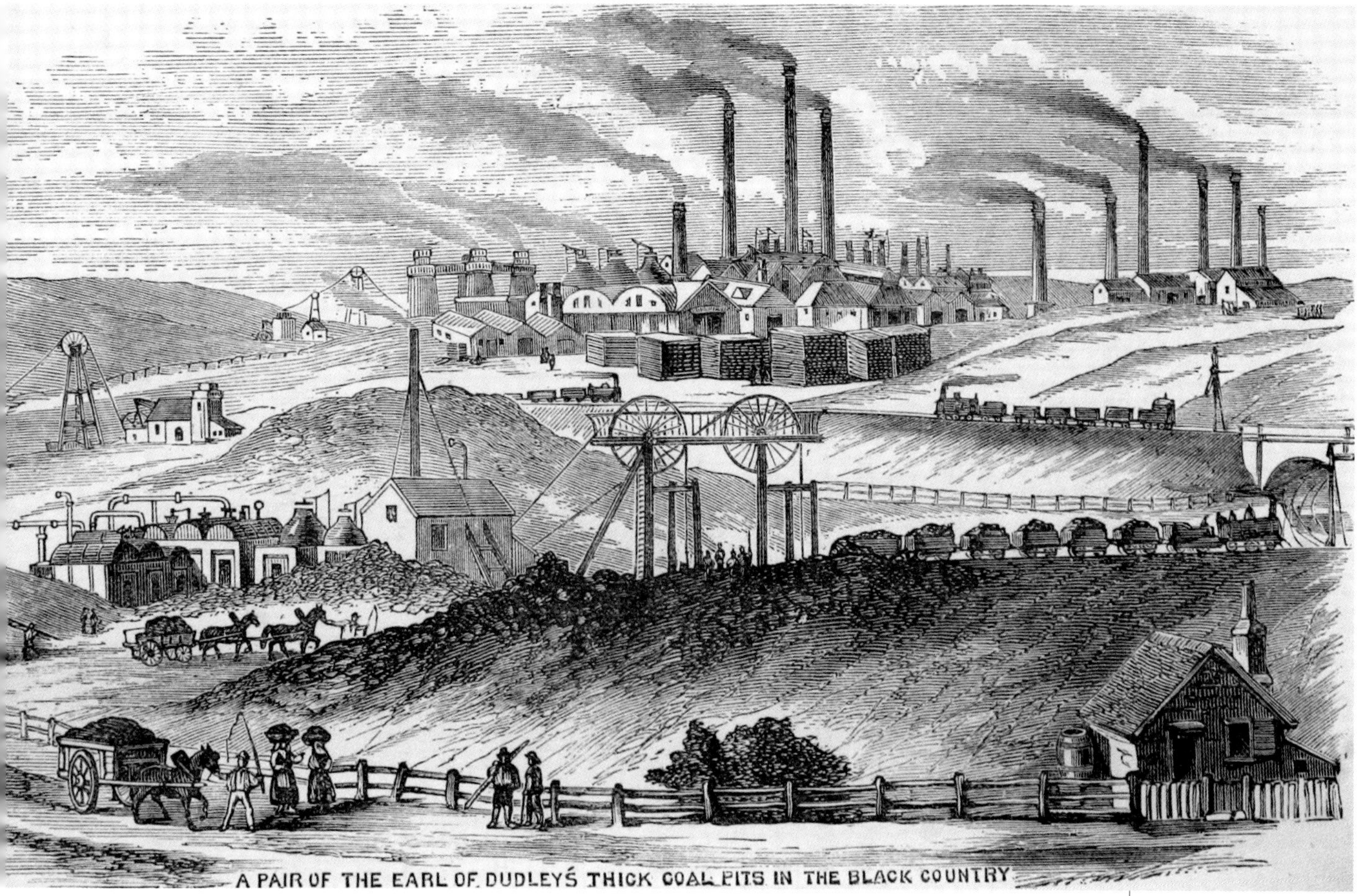

A PAIR OF THE EARL OF DUDLEY'S THICK COAL PITS IN THE BLACK COUNTRY

A contemporary print of the Earl of Dudley's coal pits. (Author's collection)

Aynho, 24 June 1961. A Hall class 4-6-0 speeds under the flyover carrying the up line of the direct Paddington, High Wycombe and Beaconsfield line, with the 8.05am Portsmouth to Birmingham Snow Hill holiday train. (Author)

A container train from Southampton to the West Midlands passing through the long closed but still intact Aynho station in the summer of 2015. It is about to pass under the distant flyover seen in the previous picture. Container trains, with their colourful cargoes and exotic ownerships from the far corners of the planet, are the present-day equivalent of the private owner coal wagons of long ago. (Author)

An LNER Robinson-designed 4-4-0 of Great Central Railway origin heads towards Banbury with the Ports to Ports Newcastle to Swansea express, c1931. The train is composed of six GWR Collett era corridor carriages, the fifth one a restaurant car, a van at the rear. (Author's collection)

No.5013 *Abergavenny Castle* accelerates away from Banbury on 8 August 1959 with the 10.8 am Newcastle to Swansea express. (Author)

The GWR and the Great Central Railway quickly came to an agreement over suburban traffic out of Paddington and Marylebone serving the affluent Buckinghamshire towns and countryside. Patronage was not great in the early years of the twentieth century, but has since steadily increased. GCR 4-4-2T No.359 has just arrived at High Wycombe with a train of neat, up to date suburban carriages, c1910. (Author's collection)

GWR 2-6-2T No.6167 at Aylesbury with a stopping train from Paddington c1960. (Author's collection)

A GWR Churchward designed 28xx 2-8-0 arrives at High Wycombe with a freight train from the London direction c1946, Note the rake of LNER Gresley designed suburban carriages in the sidings. (Author's collection)

The elderly Churchward Moguls continued to work holiday passenger traffic into the early 1960s. No.6313 has charge of a Weymouth to Wolverhampton train of eleven Maunsell era, green liveried carriages which is being held at the signals outside Banbury Saturday, 8 August 1959. Judging by the number of heads poking out of windows all along the train it has been waiting there for some time. (Author)

Canals are never far away from the Paddington to Wolverhampton route; here is the Oxford canal on 4 September 2022, less than 100 yards from Banbury station. (Author)

A typical quiet moment at Marylebone station on a Sunday lunchtime in October 2023. A Chiltern Railways train from Birmingham Moor Street has arrived with the DVT (driving van trailer) at the buffer stops; it is not yet time for passengers to board for the return journey. (Author)

Chapter 4
BANBURY TO BIRMINGHAM

It always seemed surprising that our Shrewsbury bound express, on leaving Leamington Spa accelerated straight past Warwick. Surely the county town must have been the most important place in the entire county? After all its great glory was 'England's finest castle'. I think one can just see Warwick Castle from the train, although maybe the trees now block the view since I last looked.

The GWR built Leamington Spa in 1938/9, a splendid and much needed new Art Deco station. The LMS also had a station there, Leamington Spa Avenue, of LNWR origin which was served by two lines from the east off the West Coast main line, from Roade and from Rugby. Both lines and the station are gone but the line which continued on to Coventry survives and has become an important trunk route, even though inadvisedly singled in places, and is served by Cross Country services.

The southern approach to Birmingham was markedly different to that from the north, Widney Manor, Solihull, and Acocks Green. Busy enough to warrant four tracks at Tyseley, through nice sounding and nice in reality suburbs adorned with tree lined avenues such as the line from the also very nice sounding Yardley Wood, Whitlock's End – surely a name from the Archers – Henley-in-Arden and – biggest prize of all – Stratford-upon-Avon. This brought through expresses from Bristol and the West Country, in direct competition with the Midland Railway route. This latter tended to attract the most business, not least because they owned the entire line between Bristol and Birmingham, the GWR having to settle for running powers between Bristol and Cheltenham.

The GWR missed a trick there, allowing the Midland Railway into the very heart of their territory. Two standard gauge companies, the Bristol and Gloucester and the Gloucester and Birmingham, had come into existence in the 1840s. The Bristol and Gloucester was on the point of switching to broad gauge and if the Gloucester and Birmingham had done likewise that would have brought the broad gauge into the heart of Birmingham. In the event, the year-old Midland Railway, which had only just arrived in Birmingham from the north, quickly made an offer, caught the GWR unawares and so, absorbing the complete route between Bristol and Birmingham, gained the day. Together the LNWR and the Midland constructed Birmingham New Street station whilst the GWR built the nearby, and rather more impressive, Snow Hill.

Eventually British Railways decided that there was one route too many and the through GWR one was closed south of Stratford-upon-Avon. Part of this has been reopened and is operated by the Gloucestershire and Warwickshire Railway between Broadway and Cheltenham Racecourse.

Stratford-upon-Avon being such a tourist attraction it has always seemed rather extraordinary that a direct railway service between it and London has virtually never been considered viable. During the Festival of Britain year of 1951, the Western Region laid on a through named train, complete with a Britannia pacific on the cover of the leaflet publicising it, but this was very much a one off. Today, Chiltern Railways provide a mostly hourly service, but with one change at Leamington Spa, taking just over 2 hours for the 81 miles to and from Marylebone. There have always been through trains between Stratford-upon-Avon and Birmingham; indeed nowadays Shakespeare's birthplace is a terminus, not just for regular diesel trains, but in summer for steam hauled trains in the charge of an immaculate GWR 4-6-0 based at Tyseley depot.

Birmingham was, of course, a highly desirable location for any ambitious main line railway and the competition for the traffic to and from London brought the GWR and the LNWR once again snapping and snarling at each other; although there were also times when they co-operated. The story, particularly of how the GWR reached Birmingham and beyond, is complicated one involving all sorts of wheeling and dealing which readers of a nervous disposition might like to pass over, for it involves characters who were definitely economical with the truth. The LNWR, formed in 1847, had the advantage of the shorter distance between London and Birmingham and it would not be until the opening years of the twentieth century that the GWR achieved near parity .

In 1845, the GWR promoted a line from Oxford to Rugby, the latter town at this time being the principal gateway to much of the rest of the developing railway network. Not something that pleased the London and Birmingham (shortly to be the LNWR). There would be a junction at Fenny Compton to Birmingham. But the LNWR was adamant that the GWR would never be allowed as far as Rugby and the GWR admitted defeat. Although, as it happened, it did eventually reach Rugby some 50 years later by way of the GCR.

Birmingham, back in the 1950s, never quite seemed to have the architectural style and pizazz of cities such as Liverpool or Bristol, which was probably an ill-informed opinion, and certainly on becoming better acquainted, one has recanted. Historically and geographically, Birmingham has not followed the general pattern of evolving great settlements. It, for instance, can boast no river of any significance although, to make up for this, it is the centre of the UK canal network and, indeed outscores Venice in terms of mileage – if not gondoliers. There were no great lords of the manor, and its industrial base was diverse, it not being over dependent upon any one particular product. Much of this work was carried on in people's homes or in small rather than huge work places, which has ensured the survival of many, notably the jewellery quarter, which still thrives. Much associated with the motor industry, it has also supplied a good deal of the planet with railway rolling stock.

Enter the OWW, disparagingly known as the Old Worse and Worse; or officially the Oxford, Worcester and Wolverhampton. As O.S. Nock explained in his *Great Western Railway in the 19th Century* 'for 18 years … the O.W.W. was in the hands, successively, of two very strong, resolute, and entirely self-seeking men, who thought from first to last of the financial gains to be derived from each successive transaction, and little or nothing about the job of running a railway.' These were Francis Rufford, a Stourbridge banker, 'a man of straw' and John Parson, a London banker who succeeded Rufford as chairman of the OWW and 'six months later went bankrupt'. Eventually, in 1864, the GWR acquired the OWW and its route through the Black Country to Wolverhampton and beyond was secure. Ten years earlier, the GWR's line from Birmingham to Wolverhampton had been completed; the latter would become the northern extremity of the broad gauge.

Wolverhampton plays an enormous part in GWR history, not least because its works here built many hundreds of locomotives, the last of which was only taken out of service in the 1960s, and it might well have supplanted Swindon as the company's principal manufactory. It was where expresses from Paddington changed engines, and there was always a fleet of Kings shedded here, right through to 1962 when the entire class was taken out of service, supplanted by the Swindon-designed class 52 Western diesel hydraulics. The majority of these was built at Swindon but some at Crewe – a sneaky attempt to score a last blow by the old enemy?

Preserved Castle class 4-6-0 No.5043 'Earl of Mount Edgcumbe' in the deep cutting north of Hatton with the returning 'Shakespeare Express,' September, 2023. (Author)

Leamington Spa station was completely rebuilt in the restrained but attractive Art Deco style in the late 1930s. A September 2023 view. (Author)

No. 6029 *King Edward VIII* heads through Solihull with a Birkenhead to Paddington express, 1960. (Mensing)

Worcester depot c1940. In the foreground are 0-4-2Ts Nos 5808 and 4818, a 2251 0-6-0 and in the distance a 57xx pannier tank. (Author's collection)

Birmingham had the fourth largest tram system in the United Kingdom, opening in 1904 and closing in 1953. The system had the unusual gauge of 3ft 6in This is one of the more modern trams, with upper decks enclosed, which wasn't the norm in Birmingham. Like most trams, Birmingham's were solidly built but looked very dated against the diesel double deck which in Birmingham was built to a particularly high specification. The location is outside New Street station, where once again trams have become part of the scene. (Author's collection)

The introduction of Class 52 Western diesel hydraulics in 1962 was the Great Western Regions' last attempt to retain its separation from the allegedly dead hand of BR standardisation. It had a German power unit and was styled by Mischa Black of Design Research Associates and developed a big enthusiast following. They were short lived, all being withdrawn by the end of 1977, although several have been preserved. They began their careers on the Paddington to Birkenhead route and ended them plying their trade between Paddington and Birmingham on the original route by way of Reading and Oxford. Here on 10 April 1976 No.1030 *Western Musketeer*, the centre of attention, is leaving Leamington Spa on a northbound working. (Author's collection)

No.6019 *King Henry V* speeds through the south Birmingham suburbs with a Paddington express, 1960. (Mensing)

The driver of No.5051 *Drysllwyn Castle* (sometimes known as *Earl Bathurst*) takes his ease, Tyseley in 1995. (Author)

Three 4-6-0s being prepared for royal duties at Tyseley 20 April 1950. From left to right No.7001 *Sir James Milne*, No.5000 *Launceston Castle* and No.4980' *Wrottesley Hall*. (Author's collection)

The Great Western Society's train of preserved GWR design carriages hauled by No,.5900 *Hinderton Hall* and No,7808 *Cookham Manor*, at Tyseley preservation centre in October 1978. *Hinderton Hall* had been rescued from Barry scrapyard and only just restored to steam whilst No.7808 was the only one of its class bought straight out of service. (Author)

To reach Birmingham New Street trains from the original Great Western route had to make a series of very sharp bends, first at Bordesley Junction, then St. Andrew's Junction and finally Grand Junction. Here, in 1976, two such trains, both class 52 Western hauled, pass at the latter. (Author)

The preserved one and only BR class 8P pacific, No.71000 *Duke of Gloucester* accelerates away from Banbury in 1991. (Author)

In the steam era, four tracks were needed such was the traffic passing through Solihull. Now reduced to two the route on the southern approach to Birmingham nevertheless remains very busy. A Chiltern Railways class 68 is at the head of a Marylebone to Birmingham Moor Street express in the summer of 2014, approaching Solihull, squeezed in between slower, stopping trains. Virtually brand new, No.68012 was delivered earlier that year, one of six Swiss built locomotives of the Eurolight design and capable of a top speed of 100mph. The Marylebone to Birmingham route is now a serious alternative to that from Euston. (Author)

Once, Toddington was on the through GWR route from the West Country to the Midlands. This closed in October 1976. In 1981, the Gloucestershire and Warwickshire Railway came into being with the aim of re-opening at least part of the line. This it has done and in May 2022 Churchward designed 2-8-0T No.4270 has arrived at Toddington. (Author)

An almost perfect recreation of the GWR scene in the last years before nationalisation, although it is actually 1988. No.5900 *Hinderton Hall* and No.7808 *Cookham Manor* pause at Henley in Arden with the Great Western Society's Vintage Train on its way from Paddington to Tyseley. (Author)

Chapter 5
BIRMINGHAM

There can be few great railway stations which have been razed to the ground, only to be resurrected later by those who realized what a foolish act had been committed. Such is the history of Birmingham Snow Hill. And with it we must include, at the other end of the tunnel from Snow Hill, Birmingham Moor Street, which began life as a minor adjunct, serving chiefly suburban services but eventually was revived, modernised but in a perfectly realised retro style, acquiring various prizes in the process, and now handles both local and long distance traffic.

Snow Hill was opened in 1852, but almost immediately began to expand both in its acquiring of different appellations – first being Great Charles Street, then Livery Street before settling on Snow Hill in 1858 – and size. Still not satisfied, at the opening of a new century, a completely, very grand edifice was erected between 1906 and 1912 with spacious waiting and refreshment rooms and all the usual offices as well as four main island platform faces, each long enough to accommodate two full length passenger trains, kept apart by scissors crossovers midway. There were also four bay platforms at the north end of the station. However, it was around this time that the hotel was closed, clients complaining of being kept awake by the constant racket of goods trains passing beneath; don't know what else they expected, but this was always a problem with railway hotels.

There was an inevitable air of excitement on any journey approaching Birmingham from the south in steam days, the train slowing on the approach to the tunnel, before emerging into the daylight beyond and drawing up in the station. Amidst the smoke, steam, various shadowy figures and general gloom inside the tunnel, there was also a rhythmical clattering for which there seemed to be no obvious explanation.. Then, one evening, listening to Reginald Gardiner giving one of his regular, popular monologues on the BBC light programme, it all became clear. Reginald had a cultured, BBC type voice so one knew at once that he was to be believed. Having recently passed through Snow Hill tunnel, he informed the listening public that he had concluded there was a piece of tin which stuck out from the tunnel wall and was struck by every passing train, and there were men inside the tunnel, leaning, as he explained 'on shovels, pickaxes, golf clubs, whatever was available', whose job it was to push the tin back in position after each train had passed;. although he didn't know why. Reginald's monologue ended with the farewell, 'Back to the Asylum.'

Originally a classical actor, Reginald became a great success in Hollywood in the 1930s, starring with Laurel and Hardy, and Charlie Chaplin. His career continued after the war in America, and here in films and television and the monologue became a BBC radio favourite. It is still available on the internet.

The tunnel consisted of just two tracks. This was always something of a bottleneck but the cost of boring through to add two more was said to be prohibitive so a new suburban station, Moor Street, was built up against the tunnel to accommodate traffic from the Stratford-upon-Avon, Solihull and Leamington Spa direction.

In 1964, 7.5million passengers passed through Snow Hill each year, 10.2million through New Street, big business by any standards. Steam on the route to Paddington had ended in 1962, the powerful, charismatic 2,700hp Western diesel-hydraulic locomotives taking over from the Kings, all of which were withdrawn that year, although three were to survive into preservation.

In 1967 electrification of the LMR - formerly LNWR – route between Euston and Birmingham New Street was completed and although few saw it at the time, this was the death knell of both Snow Hill and the GWR route between England's two largest cities; although even fewer foretold their eventual resurrection. With electrification of the New Street route to Euston, the old GWR Paddington to Birkenhead route was deemed redundant and closed in March, 1967. Many other services out of Snow Hill were diverted to New Street, so that the 181 departures from Snow Hill on 3 March that year were by the following Monday reduced to seventy-four. The notion that Snow Hill might eventually disappear from the railway map began to take hold. Gradually, the number of services using the station declined. By 1968 there were no trains between Shrewsbury and Snow Hill, the last train passed through the tunnel on 2 March 1968, after which all services to and from the south either terminated at Moor Street or New Street. By the end of that year there were only ten departures on weekdays, four by a single car DMU (bubblecar) to Langley Green, and six to Wolverhampton Low Level. Snow Hill had become 'the largest unmanned halt in the country', and in March 1972 it finally closed to all traffic.

It became an empty shell, almost the final indignity being its conversion to a car park, before demolition left a vast, empty space in the heart of the city.

But times and attitudes were changing. In 1972, Birmingham architect Douglas Hickman wrote, 'Birmingham people are used to changing circumstances but are human enough to want the old landmarks which gave them a sense of belonging.' These sentiments struck a chord.

The Bull Ring, adjoining Moor Street, was being totally rebuilt in the 1990s, and a grand prix style motor race was proposed around it – what joy for the shoppers. Birmingham had attracted the epithet Motorway City, which once might have suggested a brave new world, but others saw it as playing to the whims of petrolheads rather than the citizens. The West Midland Passenger Transport Authority understood the value of cross city rail services, which could only be fully achieved by re-opening the tunnel under the city and bringing Snow Hill back into use.

In 1984, £7million was raised for the first stage of the Snow Hill/Moor Street redevelopment, the line through the tunnel was opened for walkers for just one day before trains resumed, and an astonishing 13,000 people took the opportunity to do so, raising over £8,000 for charity.

Moor Street was still in use and listed, but neglected and shabby and also slated for demolition. Adrian Shooter, the Chairman of Chiltern Railways, which by then was the station's operator, realising what might be possible, enthusiastically backed plans to remove functional 1980s additions, replace them with genuine steam era canopies and generally return Moor Street to as near genuine GWR condition as possible whilst improving and sensitively modernising passenger facilities, and, for the first time, adding platforms on the through lines to Snow Hill station, which itself re-opened, in a very modest form, in 1987. A GWR-pattern water tower was built for the use of preserved steam locomotives which regularly frequent Moor Street, particularly the weekend summertime 'Shakespeare Express' to and from Stratford-upon-Avon, the station is resplendent in GWR colours and GWR style lettering. Fittingly, Moor Street received the Railway Heritage Trust Conservation Award in 2004 and the Birmingham Civic Renaissance Award in 2006.. It is an example to all cities of how to honour Victorian icons whilst making them fit for twenty-first century purpose.

The locomotives which operate the 'Shakespeare Express', Halls and Castles, including No.7029 *Clun Castle* which hauled the last regular steam train out of Paddington on 11 June 1965, are based at Tyseley, where the Stratford-upon-Avon and Leamington Spa lines diverge.. A large depot was opened here by the GWR in 1908 and was home to over 100 steam locomotives in GW and BR days. Today, the depot is in two parts, the preservation centre and a stabling and maintenance facility used by West Midland diesel trains.

The eastern approach to New Street station actually passes right under the southern portal of Snow Hill tunnel so that it is possible to stand on Moor Street station and watch trains on their subterranean journey. New Street station, which never rivalled the grandeur of Snow Hill in its heyday, itself has been totally and most spectacularly rebuilt, at least above platform level, although neither the present day Snow Hill nor New Street platforms are environments anyone other than an avid train spotter, would want to linger longer than absolutely necessary. Oh for a grand arched roof. There may be a lot less smoke and steam but there are diesel fumes aplenty.

Snow Hill reopened on 5 October 1987. Nothing like as grand as its predecessor, just four platforms with a modest concourse up above, although a big expansion is planned, all in the heart of the city. It has, inevitably, become vital to the movement of people around the city and beyond, not least since the return of trams, which have taken over the track bed where once Kings and 2-6-2Ts plied their trade between Birmingham and Wolverhampton.

A pre-1914 Birmingham Snow Hill scene. Dominating the picture is one of a class of four Armstrong 4-4-0s with 6ft 8in driving wheels. Nominally dating from 1894 but extensively rebuilt more than once they moved to the Wolverhampton Division in 1909/1911. The Armstrong is piloting a Bulldog 4-4-0 and the train is likely to be a through one from Paddington to Birkenhead. (Author's collection)

BIRMINGHAM • 55

Postcards in full colour were produced of the almost new Snow Hill station, situated on Colmore Road, just across the street from the cathedral, with an equally new and impressive double deck electric tram about to turn out of Livery Street. (Author's collection)

Deep in the gloom of Snow Hill tunnel the carriages lights of a passing train illuminate the walls. (Author)

No.5088 *LLanthony Abbey* tender piled high with coal, seems to be anxious to get away with its southbound stopping train, 19 September 1952. Built as a Star class engine in January 1923, it was rebuilt as a Castle in the late 1930s and was eventually withdrawn in December 1962. (Author's collection)

Over at New Street c1923, an elderly Webb LNWR saddle tank shunts a couple of carriages, one in LNWR livery the other in Midland/LMS red. (LNWR Society)

In May 1999, trams returned to the streets of Birmingham, initially from beside Snow Hill station, then along the trackbed of the abandoned Snow Hill to Wolverhampton Low level railway station which, of course, had been part of the Paddington to Birkenhead main line. The original trams were replaced in 2014 as the system expanded. One of the original trams is seen at the cramped Snow Hill tram platform. (Author)

Birmingham was and is the centre of the UK canal network. Today, as in their Victorian heyday, there are more canals in the Birmingham than in Venice; but rather fewer gondoliers. (Author's collection)

After demolition, Snow Hill station, Birmingham was a great open space, a temporary car park; a glimpse of its former greatness is a surviving piece of tiling from a refreshment room. Rather like an archaeologist's excavation of a Roman bath house floor. (Author)

No.4865 *Rood Ashton Hall* arriving at Moor Street from Stratford upon Avon in the summer of 2014. (Author)

Crowds head for a Chiltern Railways class 168 DMU at the terminus platforms, Moor Street in 2013. (Author)

Preserved Castle class 4-6-0 *Defiant* based at Tyseley in steam there in 1995. As well as a preservation centre Tyseley is also a depot providing a home for many of the DMUs which work in the Birmingham area. (Author)

By great good fortune the original arch at Curzon Street has survived. It began as the Birmingham terminus of the line from London, then for most of its life served as a freight depot. The planning of the high-speed line (HS2) from London to Birmingham and beyond recognised its importance and one day, despite all the delays, it will resume former glories. (Author)

CHAPTER 6
WEST MIDLANDS

The story of the railways of the West Midlands is interwoven with that of the street tram system and one place in particular, Dudley, features throughout that story. Trams returned to the streets of Birmingham on 6 December 2015 when they began operating between Birmingham and Wolverhampton, for much of their route along what had been part of the GWR Paddington to Birkenhead main line . Once there was a huge tram network extending all over the West Midlands, but its heyday was brief, from around the beginning of the twentieth century until the mid-1920s, when decline set in. The very last system, that of Birmingham Corporation, was closed in 1953.

Dudley was served not only by trams, but also the LNWR and the GWR. Dudley was right at the very centre of the Industrial Revolution, its citizens paying the price of pioneers who laboured in the operation of canals and railways, down coal mines, beside blast furnaces, hammering away at chain making, taking part in virtually every form of manufacturing known to man and finding themselves living in 1851 in 'the most unhealthy place in the United Kingdom'.

More than 100 years later, their ancestors were able to gain some benefit from all this toil and exploitation when in 1978 the Black Country Museum was opened in Dudley, bringing to life the good, the bad and the ugly days of the not so long ago, some within the memory of its oldest inhabitants. Once again it was possible to ride on a tram or a trolleybus, experience an all through school where pupils went out into the world to earn a living at the age of fourteen, to propel by the power of their legs a narrow boat through the second longest canal tunnel in the country, go down a coal mine accompanied by a miner with the broadest of local accents, indulge in coconut shies and swing boats at a travelling fair, learn how on a Friday night the youngest of the family would be last into the now tepid and less than inviting water in the tin bath in front of the kitchen fire, and so much more.

Dudley GWR station was on the line from Stourbridge to Wolverhampton. Adjoining it was the LNWR line, which diverted a short distance to the south of the GWR one and continued on to Dudley Port High and Low Level stations, the former situated on the LNWR main line between Birmingham New Street and Wolverhampton High Level. The latter went on to Wednesbury, but there were other junctions and connections in the vicinity between the two companies. All of these are now gone, except for Dudley Port, the High Level suffix no longer needed. Dudley station closed to passengers in 1964, although freight continued until 1993. In the meantime, the short-lived Dudley Freightliner Terminal opened on the station site in 1967 and closed 22 years later.

The West Midlands has become an enthusiastic proponent of the tram, often making use of underused and abandoned rail lines and in 2017 work began on the extension of the West Midlands Metro from Wolverhampton by way of Wednesbury via Dudley to Merry Hill, which also included reviving a parallel freight line through Dudley. In February 2021, surveys began along the route to check on the many old mineshafts in the vicinity.

Wolverhampton may not be the prettiest of cities, although those with a nose for the unexpected will find it quirky, fascinating and of great historic significance, especially in relation to the GWR and industry generally. It has always been a frontier town – it only became a city in the year 2000. One doesn't mean on account of the gunslingers roaming its streets, never numerous and now few and far between, but because it was where a King class locomotive handed over to a less powerful cousin, heavy industry ended, to be replaced by a land devoted to agriculture with the high hills and mountains of the Welsh Marches soon appearing on the distant, western horizon.

As far as the GWR was concerned, Wolverhampton ranked second only to Swindon in terms of locomotive development, at least until the end of the broad gauge in 1892, and one family name, Armstrong, ranked along with those of Gooch, Dean and Churchward in terms of GWR significance . Wolverhampton railway works was opened by the Shrewsbury and Birmingham Railway in 1849. Five years later, this became the workshop of the northern division of the GWR which was then rebuilt and enlarged under the expert guidance of the engineer in charge, Joseph Armstrong.

No-one could have been better qualified than Joseph. His pedigree for the post was impeccable. He attended the same school in Newcastle as Robert Stephenson, had watched Stephenson engines at work on the Stockton and Darlington Railway, was employed as a driver on the Liverpool and Manchester Railway, acquiring skill and experience in this newest of all technologies and eventually was put in charge of the locomotives of the Shrewsbury and Chester Railway. This was not a simple task, for hardly two locomotives were alike. For his first few years in charge sorting out this motley collection took up all his extensive energies. It was not until 1859 when the first Armstrong-designed ones emerged from the Stafford Road works at Wolverhampton.

Gooch resigned in 1864, eventually becoming chairman of the Great Western Railway, and Armstrong was moved to Swindon. He was not only a talented engineer but a man of conviction and principle, a Methodist lay preacher who encouraged all forms of Christianity and took the welfare of his men and their families seriously. What we would now term a workaholic, Joseph Armstrong's workmates were concerned for his health. Persuaded to take a break and recuperate in Scotland, he died of a heart attack in 1877 on his way there.

He had been succeeded at Wolverhampton by his younger brother, George. George was an independent, free-thinking man who only gave orders, did not take them. Under him Wolverhampton became semi-independent of Swindon, painting its locomotives in its own shade of green, and largely concentrating on a succession of 0-4-2 and 0-6-0 tank engines which served the GWR well, many modified and rebuilt, surviving almost until the end of steam. O.S. Nock, writing in 1930, commented that ' those old Armstrong Metro 2-4-0s were an absolutely classic example of precision suburban train performance. It did not matter which of them was on the job; the running was as regular as clockwork.' There was even serious discussion at one time of Wolverhampton succeeding Swindon as the principal manufacturing base of the GWR. One of George Armstrong's tasks involved accompanying Queen Victoria on the royal train as far as Bushbury, Wolverhampton, where it was handed over to the LNWR. He is said to have done this over 120 times. George retired at the age of 75 in 1897. New construction ceased at Wolverhampton in 1908, but heavy repairs continued until the end of steam. The last Wolverhampton-built locomotive, 2-6-2T No.4507, was withdrawn in 1963. The last repaired there, 2-8-0 No.2859, has been preserved.

A long lived former LNWR 0-8-0 freight locomotive shunts near the junction of the former GWR and LNWR routes at Dudley in 1958. These seemingly indestructible old engines soldiered on into the West Midlands well into the diesel era, the last not being withdrawn until 1962. (Author's collection)

One of the third generation of Birmingham trams approaching West Bromwich southbound in September 2022 along the track bed of the former GWR main line. (Author)

LMS pacific No.6201 *Princess Elizabeth*, at the Liverpool and Manchester Railway 150 celebrations. It regularly worked the Merseyside Express in pre-preservation days. (Author)

Paddington Underground station, c1891 from a contemporary child's picture book. (Author's collection)

A Watford to Euston electric train passing over the Regent's Canal on the approach to its destination in October 2023. (Author)

Preserved GWR 4-4-0s No. 9017 and No. 3440 *City of Truro* haul a Llangollen bound train near the Horseshoe Falls in the spring of 2009. (Author)

No.6023 *King Edward II* is one of the three preserved King class 4-6-0s. (Author)

No.7029 *Clun Castle* accelerates the very last regular steam hauled passenger train out of Paddington on 11 June 1965. (Author)

Shrewsbury station in the spring of 2021. In the distance is the Severn Bridge Junction signal box, the largest mechanical signal box in the world. (Author)

The Mersey ferry, *Snowdrop*, sporting Sir Peter Blake's dazzling livery, sets off from Liverpool Pierhead, 2022. (Author)

Broad Street, Oxford in November 2015. (Author)

A replica of the 1847 vintage Broad Gauge 4-2-2 locomotive *Iron Duke,* seen at an Old Oak Common open day. (Author)

No.70000 *Britannia*, the very first of 999 British Railways standard steam locomotives, having departed at Crewe is seen curving out of Chester bound for Shrewsbury in 1981.
(Author)

A 1930s LMS poster.
(Author's collection)

A late 1930s 'GWR sunshine' window first class carriage. (Author)

Coal empties bound for South Wales seen from Stokesay Castle, on the Marches main line in 2002. (Author)

At almost the same venue a First Great Western class 175 DMU is heading north in June, 2022. (Author)

Looking north from Shrewsbury Castle in 1977. (Author)

Rock Ferry in 1970. An English Electric class 40s is coming up from the docks whilst an LMS class 503 EMU heads for Liverpool. (Author)

A summer 2012 scene at Moor Street Birmingham. (Author)

Coal empties hauled by a class 66 grind their way up passing the oldest working iron works in the world on the ascent from the Severn Valley at Ironbridge, 2009. (Author)

Birmingham Broad Street, heart of canal country. (Author)

Preserved GWR diesel railcar No. 22, comes to a halt at Arley station on the Severn Valley Railway in 1980. (Whetmath)

Replica LNER A1 pacific *Tornado*, leaving Bridgnorth in 2010. (Author)

Midland Red S16 of 1964 and Birmingham Guy Arab No.2976 of 1953 meet. (Author)

Birmingham Moor Street station in 2009. (Author)

Woodside ferry c2010 with restored Birkenhead built tram No.20. (Author)

WEST MIDLANDS • 65

The diesel era at Dudley c1975 with a 08 at work. (Author's collection)

On the canal at the Black Country Museum. (Author)

GWR diesel rail car No.8 on a local working to Birmingham Snow Hill from Dudley, c 1947. (Author's collection)

Joseph Armstrong. (Author's collection)

The pioneer King class 4-6-0, No.6000 *King George V*, tender piled high with coal, at Stafford Road depot, Wolverhampton, about to take over a Paddington bound express, 1961. Alongside is a Castle. (Author's collection)

Class 90 Electric locomotive, Wolverhampton High Level, 1999. (Author)

Wolverhampton Low Level, 16 June, 1957. On the left preserved GWR 4-4-0 No.3440 City of Truro with an excursion to Swindon, on the right 2-6-2T No.5106 with a stopping train for Strourbridge Junction and Worcester. (Author's collection)

A sad and rather extraordinary photograph of Wolverhampton Low Level station sometime after all traffic had ended on 12 June 1981. (Author)

A Shrewsbury bound class 158 heads out of Wolverhampton over the canal, April 2023. (Author)

Chapter 7
SHROPSHIRE BRANCH LINES

The River Severn was enormously important at the beginning of the Industrial Revolution, but tidal fluctuations meant the journey into the Bristol Channel could be delayed by weeks. Canal building began in the 1780s, but the tub boats used on the canals were much shorter than the standard narrow boats which became the norm elsewhere.

Railways were the answer, and these were provided by the GWR and the LNWR. The GWR (as it would become) line from Shrewsbury, down the Severn Valley to Ironbridge, Bridgnorth, Bewdley and Stourport on Severn, opened in 1862, and was the principal one in the area, although it was very much a branch line and virtually all single track. It was one of several that served the various industrial sites bounded by Wellington, Ironbridge, Coalport and Shifnal. In 1854, the GWR had opened a line from Wellington, through Horsehay to Coalbrookdale, to link with the Shrewsbury to Bridgnorth line at Buildwas Junction, while in 1857 the first section of a LNWR line from the Wellington to Stafford branch by way of Oakengates to Madeley and Coalport began. There was also a network of privately owned lines serving collieries and other works, the Lilleshall Colliery, for instance, operated over 50 miles of track.

These lines were very much freight oriented. Madeley Court saw just two passenger trains at day, and these ceased in 1925. However, evening summer excursions to bathing places on the Severn were popular. Passenger trains on the LNWR line to Coalport, which was also served by the GWR's Shrewsbury to Bridgnorth route, never warranted more than a couple of carriages, sometimes only one, and after the First World War the buses of the Midland Red, a company which covered a greater area of England than any other, often proved more convenient than the train.

The GWR line from Wellington and Madeley Junction, once past Coalbrookdale and Ironbridge, changed character completely and became deeply rural. The only station generating much traffic was the picturesque town of Much Wenlock, after which came five more poorly-patronised stations before it joined the LNWR/GWR north to west main line at Marsh Farm Junction, immediately north of Craven Arms. Powell and Pressburger set their 1950 film, *Gone to Earth*, in and around Much Wenlock. One critic said of it, 'Where else in the whole of cinema is there a more haunting and magical evocation of the English landscape?'

The decline of the railways in south east Shropshire had begun between the two world wars and continued apace after 1945. The last coal pit, the Glanville, serving Ironbridge power station, ceased work in 1968. The line from Madeley Junction to the power station survives but is no longer used, the power station closing in November 2015. In 1965, preservationists began the process of reopening the 16 miles of the closed Severn Valley line south of Bridgnorth which eventually was reconnected with the national network at Kidderminster. It has a quite wonderful selection of locomotives and rolling stock, to say nothing of its beautiful stations. There is also a small preservation centre at Horsehay.

In its last British Railways years, the network of lines was home to an extensive collection of antiques, Webb designed tank engines of LNWR origin for instance with carriages to match, whilst the very last Dean Goods 0-6-0, No.2516, found a final working home in Shropshire, at Oswestry, before retiring into preservation at STEAM, the museum of the GWR at Swindon.

There was one other line out of Wellington. This was scarcely a branch, being a serious attempt at a drive by the GWR deep into the heart of the LNWR empire at Crewe. It even opened a depot there, Gresty Lane. Rather ironically, considering Gresty Lane was such a minor affair, diesel locomotives are still stabled nearby.

The first section of the Wellington to Crewe line, between Nantwich, on the LNWR's Shrewsbury to Crewe line, and Market Drayton, opened in October 1863, and the second and final part, from Market Drayton to Wellington, exactly four years later. Market Drayton was the largest place on the line, but it was not, in the great scheme of things, that important. In reality, passenger traffic never required trains of more than two or three carriages, hauled by not very large locomotives, 2-6-2Ts, and pannier tanks, and, in earlier days, 2-4-0s and 4-4-0s, stopping at all stations, and freight was what you would expect in an area more or less totally dependent on agriculture. But during the two world wars it became of enormous importance, carrying vast numbers of troops and large amounts of freight between Scotland, the north of England and the south coast, relieving the various stretched other routes.

There were two RAF Second World War aerodromes close to the line. There was Shawbury, also close to Hadnall on the Shrewsbury to Crewe line and one used regularly to see RAF vehicles there. Tern Hill was a little way south of Market Drayton and much traffic originated there, including a regular train from RAF Sealand, near Chester, on Saturdays bringing recruits for advanced training, the motive power, certainly in the early days, being shared by elderly 4-4-0s, GWR Bulldogs and ex LNWR Precursors.

Throughout its existence, the Crewe to Wellington line was used when long distance passenger trains had to be diverted from their regular route. Then just about any form of motive power including GWR Castles and Stanier pacifics might appear, and for a brief period the 'Pines Express' passed this way regularly in the

A replica of Trevithick's steam locomotive of 1802, which was commissioned by the iron works at Coalbrookdale, now part of the Ironbridge Gorge Museum Trust . None of Trevithick's locomotives has survived; this one is based on a drawing in the Science Museum. Cornish born Richard Trevithick was the first to show the power of steam when applied to hauling loads along iron rails. (Author)

years 1962/3, usually hauled by a Warship diesel hydraulic or an English Electric class 40, although steam in the shape of a Castle from Stafford Road Shed appeared at least once.

I never travelled upon the line, but I did persuade Uncle Frank to cycle with me over to Hodnet one Sunday morning where we watched a Crab 2-6-0 shunting in the goods yard. Whilst on a visit in 1956 I happened to be riding my BSA Bantham under a bridge near Crudgington when the last Star shedded at Stafford Road shed, No.4061, *Glastonbury Abbey*, steamed past hauling a Crewe bound freight. The line closed completely on 8 May 1967.

The world's first iron bridge, a picture taken in 1954. Although recognised for what it was, this was some years before the concept of the Ironbridge Gorge museum, really a series of museums, centred around the River Severn and the birthplace of the Industrial Revolution, came into being. (Author)

Star class 4-6-0 No.4053 *Princess Alexandra* pulls out of Wellington in 1950 with a Paddington express. (Morrison)

Coalport Station East. Coalport, a village deep in the Severn valley near Ironbridge, could boast no less than two stations. This was the LNWR one, the terminus of a branch off the Shrewsbury to Stafford line. It opened in 1861 and closed to passengers in June 1952, to freight eight years later. . The scene is pure LNWR, despite being taken the best part of thirty years after that railway was absorbed by the LMS, the locomotive and two carriages all originating on that railway. (Author's collection)

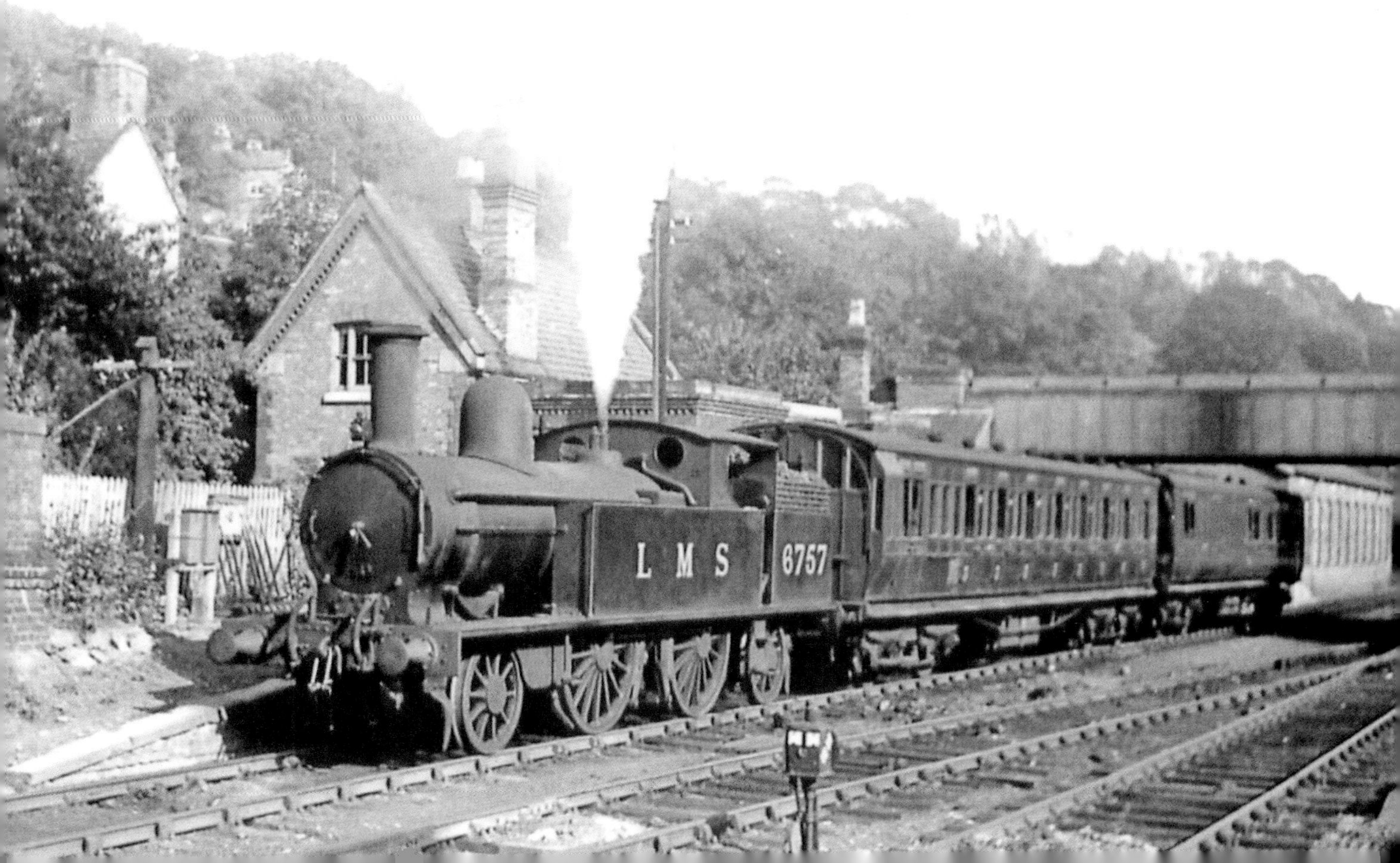

Dean Goods No.2516 on the other line which served Coalport, that from Shrewsbury to Bridgnorth and Hartlebury in 1955. By this time No.2526 was the last survivor of a long-lived class having been built in 1897. Shedded at Oswestry, it was withdrawn the following year. (Author's collection)

Following withdrawal No.2516 was brought to Swindon and spent some time in store in the company of the preserved No.4003 *Lode Star*, before restored to GWR unlined green livery. Today it can be seen at STEAM, the museum of the Great Western Railway at Swindon, where aspiring drivers can mount its footplate and imagine they are trundling off along some remote rural byway. (Author)

No.3205 at Bridgnorth in 1970. This Swindon built 0-6-0 performed the opening ceremony of the Severn Valley Railway in May 1970, No. 3205, built in 1946, is the only survivor of the Collett 2251 class 0-6-0s and worked the Severn Valley line in BR days when shedded at Shrewsbury. (Author)

The Great Western Society's train of preserved GWR design carriages passing through Wellington in 1969. (Author)

Wellington shed with a couple of pannier tanks and a 5101 2-6-2 in residence c1955. (Author's collection)

Chapter 8
WOLVERHAMPTON TO SHREWSBURY

In many respects, Shrewsbury has all the qualities of a city, and was indeed offered that status by Henry VIII, but the good town fathers declined, knowing that Henry was only making the offer in order to extract much higher taxes. Shrewsbury, although it is not very large, has many of the attributes of a city, including a university centre, a strategic position of vital importance to Wales/England links on the longest river in England and Wales, an abbey, a magnificent parish church, a great public school, and a wealth of fine architecture.

Before we reach Shrewsbury from Wolverhampton, we have to leave Staffordshire, which is not a county most of us associate with the GWR. There are eight stations between Wolverhampton and Shrewsbury, a surprisingly high number given what has happened to many wayside stations elsewhere between Paddington and Birkenhead. The first across the border is Albrighton. When gas arrived there in 1868, the gasometer was logically built beside the railway goods yard. Gasometer and goods yard are now gone but passenger traffic is buoyant. Albrighton has grown a lot in the last ten or so years, being within comfortable commuting distance of just about anywhere between Wolverhampton and Birmingham, and business has more than doubled.

When I first travelled this route, all most of us knew of Telford was that it was the name of the great civil engineer. And civil engineers were never as famous as those who built machines which moved and belched out loads of smoke and steam, such as Trevithick, George Stephenson and the like. Of late, in the last fifty years or so, Telford's star has risen to the heights so that he is recognised as being in the very first rank of those who created modern Britain. A Scotsman, like so many of the great pioneers of the Industrial Revolution, Telford, who served an apprenticeship as a stonemason, was basically self taught and in 1787, at the age of 30, became Surveyor of Public Works in Shropshire. His mark is to be found over all the county, in some forty bridges, including his first iron one, at Buildwas, which, although no longer in existence, was a huge improvement on that at Ironbridge, Then there is the fabulous Pontcysyllte Aqueduct on the Shropshire/Welsh border, a world heritage site, beside which our London to Birkenhead railway later joined it. In 1820, he became the first President of the Institution of Civil Engineers and held the post until his death in 1834. He is buried in Westminster Abbey.

When a new town was proposed in the 1960s embracing much of that part of Shropshire stretching from Ironbridge to Wellington, Dawley and district it was inevitable that it would be named Telford. Pity that Telford Central's station is a miserable, penny-pinching affair not worthy of bearing such a distinguished name. Through expresses still only stopped at Wellington, a much more substantial and pleasing piece of architecture, but as Telford, after some initial uncertainties, has grown and established itself, Telford Central now has an excellent train service. There are fifty-five trains a day to Wolverhampton, the journey taking between 19 and 30 minutes, and 57 to Shrewsbury, the journey taking 20 to 27 minutes with stops just at Oakengates and Wellington; the three stations between Wellington and Shrewsbury, Admaston Halt, Walcot and Upton Magna all closed in the summer of 1964.

Thomas Telford. (Author's collection)

One of Churchward's original County class 4-4-0s speeds along the Welsh Marshes near Church Stretton c.1918 with a North to West express. Built between 1904 and 1912 the Counties were fast, powerful, and somewhat rough riding, and did excellent work on the North to West main line until withdrawal between 1930 and 1933. (Author's collection)

Hereford on a June evening in 1956. A Hawksworth County class 4-6-0 piloting a Hall, both newly restored to lined green livery. The County was the very last new variety of GWR 4-6-0, the design of which was being prepared at Swindon as the Second World War was coming to an end. Like their 4-4-0 predecessors they were a familiar sight on the North to West main line. Shrewsbury always had several, until dieselisation wiped them out. A replica of one is under construction. (Author)

The GWR went in for big, eight coupled tank engines, they being very useful in hauling long rakes of coal wagons up and down the steep Welsh valleys. The Wall Street crash of 1929 and the subsequent decline in the coal industry meant that many of them had to find other work Some had a rear pony truck added and an enlarged bunker so that they could become fully-fledged, long-distance freight locomotives. Others, such as No.5243 seen at Hereford in the spring of 1956, migrated eastwards. (Author)

No.7211, one of the big Collett 2-8-2Ts, rebuilt from a 2-8-0T, on the approach to Colwall Tunnel, beneath the Malvern Hills, 9 August 1960. (Author's collection)

West Midland Railways No.170526 approaching Great Malvern with a train from Birmingham to Hereford in July 2022. The 170s were built at Derby, the first entering service in 1998. (Author)

No.7, one of the earlier GWR AEC/Park Royal diesel railcars heads below the Malvern Hills on its way from Worcester to Hereford, c1939. (Author's collection)

The Great Western Society Vintage Train of preserved chocolate and cream painted carriages wends its way through the Malvern Hills hauled by No.7807 *Cookham Manor* and No. 6998 *Burton Agnes Hall* in 1979. (Whetmath)

A LNWR Prince of Wales 4-6-0 stands in Hereford Station, c1920 with a northbound train. The first Prince of Wales emerged from Crewe in February 1912 and the last of this class of 245 was completed in February 1924. Some of the last survivors regularly worked the Shrewsbury to Crewe line, being shedded at Shrewsbury in the 1940s. Four lasted into British Railways days, the last taken out of service in May 1949. (LNWR Society)

Britannia class pacific No.70025 *Western Star* storms north past Little Stretton 19 September 1958 with the 11-coach 8.20am Swansea to Manchester London Road. At this time most north to west trains were still composed of pre-nationalisation carriages. The Western Region was initially allocated fifteeen Britannias and eventually the entire batch was concentrated at Canton, Cardiff. (Author)

The Joyce clock, Hereford Station. Joyce of Whitchurch, Shropshire, could lay claim to be the oldest clockmakers in the world, having been founded in the Shropshire village of Cockshutt in 1670. One hundred years later they moved to Whitchurch, from which they continued until the last member of the family, Norman Joyce, retired in 1964. Joyce clocks could and still can be found all over the LNWR system, including Liverpool Lime Street and Carnforth. (Author)

Rebuilt Royal Scot No.46106 *Gordon Highlander* heads south through the Stretton Hills on 6 September 1958 with the 9.25am Manchester London Road to Swansea. Most of the twelve carriages are of GWR origin, with a couple from the LMS and one a post-war ex LNER of Thompson design. (Author)

Train spotters at Hereford, 1984. Motive power consists of two 47s and one 37. (Author)

The North to West main line, with its attractive scenery and plenty of pathing has always been a favourite for steam excursions ever since the early 1970s, and here, in 1981, the once streamlined Stanier pacific No.46229 *Duchess of Hamilton* (it has since been re-streamlined) is seen at Craven Arms., its train extending back on to the Mid Wales line. Stanier pacifics certainly never appeared on the Mid Wales line. Note the first carriage which is a LNWR built brake composite. (Author)

Chapter 9
SHREWSBURY

The first railway to reach Shrewsbury came in from the north west, from Chester, in 1848, that from Wolverhampton and London a year later. Shrewsbury was to become, and has remained, a junction of great importance, indeed it could claim to be of international significance. To quote John Horsley Denton in his *Shrewsbury Railway Station, a Brief History*, written in May 1993, 'Many Welsh organisations held their general and committee meetings in Shrewsbury because it was the only place to which delegates could come by train from most parts of the Principality and be able to return home the same day.' Recently, the Welsh government has announced its financial support for an upgraded 'premium service' linking Holyhead and Cardiff via Shrewsbury. Indeed, in railway terms, Shrewsbury seems to have actually migrated across the border, Arriva Wales claiming ownership of all its Shropshire stations.

As a piece of railway architecture, Shrewsbury ranks right up there amongst the most distinguished. While not quite the equal of St. Pancras, it can look Brighton, Liverpool Lime Street, Wemyss Bay and such like, in the cast iron column, cornice or wherever appropriate , with a steady, unflinching gaze. Its designer was Thomas Mainwaring Penson, who did all the stations between Chester, where his practice was based, and Shrewsbury. The stone came from Grinshill, nearest station Yorton. The style Penson employed is best described as Tudor Gothic, bearing mind that it sits alongside the castle and had thus to pay homage to its neighbour, and also that the town is full of wonderful Tudor buildings, as well as equally handsome Queen Anne and Georgian examples. Shrewsbury Castle, Norman in origin, became a residence in Tudor times, passed into the keeping of the town council in 1600 and has been much restored over the years, by Thomas Telford amongst others. Although this was not on the agenda of either the Normans or Tudors, let alone Thomas Telford, it has always been an excellent vantage point for watching the comings and goings in and around the station.

Rather remarkably, the station had a complete new third storey added as the twentieth century dawned, but not on top as might have been thought logical, but down below, and this involved digging down so that the front entrance had also to be lowered by a whole storey. Much of the station extends out over the River Severn, so that trains coming in from the south, whether from the east or the west, all have to make a very slow, curving approach in order to straighten up to gain the platforms, which extend right across the river to the far bank. On departure, trains bound for the GWR route to Chester, Birkenhead and to North Wales make a more or less straight beeline past the now vanished Coton Hill marshalling yard, where as late as the 1960s cattle still arrived from the Republic of Ireland. Those bound for Crewe take yet another extremely tight bend before the steady climb past Battlefield and Hadnall Bank, a bend so tight that it was the scene of a disaster in October 1907.

The heavy overnight mail and passenger train from Crewe with LNWR and GWR carriages, some of which had begun their journey at either Glasgow, York or Liverpool, was hauled by Experiment class 4-6-0 No.2052 *Stephenson*. It should have slowed to 10mph on the approach to Shrewsbury, but careered on at what was estimated to have been 60mph across the curve, spilling carriages in a splintered pile up, causing the deaths of eighteen people, passengers, mail sorters and both driver and fireman of No.2052. Thirty-three others were injured. Inevitably, the spectacle caused a huge amount of horrified fascination and my Uncle Frank was one of many children who were taken by their parents the next day to see the wreckage. An enquiry concluded that there was nothing wrong with the brakes on the train, and that an examination of the driver's body showed no trace of drugs or alcohol. The guard, who was at the rear of the train and survived, said that as he realised the train was travelling much too fast as it approached Shrewsbury station, he applied the brakes and felt the driver do the same, but it was by then much too late. In the end, the verdict was that for some reason the driver's attention had been distracted and he was unaware of where he was until it was too late.

Being jointly owned, Shrewsbury was well provided with waiting and refreshment rooms and all the usual offices. In Edwardian times there were 1st and 2nd class refreshment rooms, but only a 3rd class waiting room (ladies). The urinals were unclassified.

Houses were demolished to make way for the station as it grew to its fullest extent, complete with overall roof, three through platforms and four more terminal ones at the west end of the station, over the river. Preceding the railway was the Shropshire Union Canal and although rail and canal survived alongside each other for several decades, inevitably the canal traffic faded away and the basin was filled in.

One could hardly wish for a more central place for a station, and that advantage must have ensured that so much of the traffic from far and near, which came in on the no fewer than nine routes which eventually connected Shrewsbury with the rest of the UK, has remained, although inevitably a good deal has been lost since the motor vehicle appeared on our roads.

Whilst on that subject Shropshire played a leading part in the relatively short period when the steam lorry was a common sight. This was due to a name, which still rings thrillingly though any Salopian proud of his or her county's industrial history, Sentinel. No steam fair worthy of the name will be without several Sentinel traction engines or lorries, and probably several. These were built at the Sentinel works set up alongside the Shrewsbury to Crewe railway line on the Whitchurch Road in 1915. Sentinel's reputation was second to none and they continued to build steam lorries longer than any other manufacturer, the last for use in the UK, the celebrated S type, being turned out in 1938, although some were built for the Argentine government in the early 1950s. Sentinel also produced diesel lorries and buses, but they had also had a long history of constructing steam powered railcars and small industrial shunting locomotives, later ones being diesel, which had widespread use. The works was taken over by Rolls Royce in 1956, producing diesel engines. The name still survives, carried on by Sentinel Manufacturing, based across the road on the Battlefield Industrial site; some of the original buildings, and particularly the main entrance with its handsome statue of a sentinel, remain.

Shrewsbury actually had a second station but only a select few ever knew of its existence. It was only a few hundred yards distant, just across the English Bridge, and was the terminus of the Shropshire and Montgomery Railway. Named the Abbey on account of the fact that it occupied part of the site of the much-reduced Medieval abbey and dominated by the still impressive surviving part, it belonged to a railway which spent much of its career in the hands of receivers and lawyers, passing in 1909 to the wonderful Colonel Holman F. Stephens. He devoted himself to reviving lost causes, in the shape of rural railway byways. The complexities of the alliances the Shropshire & Montgomery attempted to form, and sometimes managed, and the antiquity of its rolling stock were wondrous indeed, but it finally gave up carrying passengers on 6 November 1933, although some goods trains remained. The Second World War saw the army come to its rescue, taking over in June 1941 and setting up ammunition dumps along the line. A wise choice for, perhaps not surprisingly, the Luftwaffe paid little attention to Shropshire.

The Shropshire and Montgomery was absorbed by British Railways in 1948, the Army gradually reduced its presence, and the last trains ran in March 1960, after which dismantling began. The goods yard remained in use as an oil depot until July 1988, while a preservation group began restoration of the passenger station around 2005.

Shrewsbury station was designed to fit in with the distinguished architecture of the border town and thus a Tudor style, with Italianate additions was what was required. And here it is, sometime in the 1920s. (Author's collection)

Much of Shrewsbury's abbey has survived and is seen here, across the A5 road from the Abbey station. This closed to passenger traffic in November 1933 but revived during the war and freight in one form or another continued into the diesel era. In the twenty-first century, a preservation group has been established and is reviving the site. (Author)

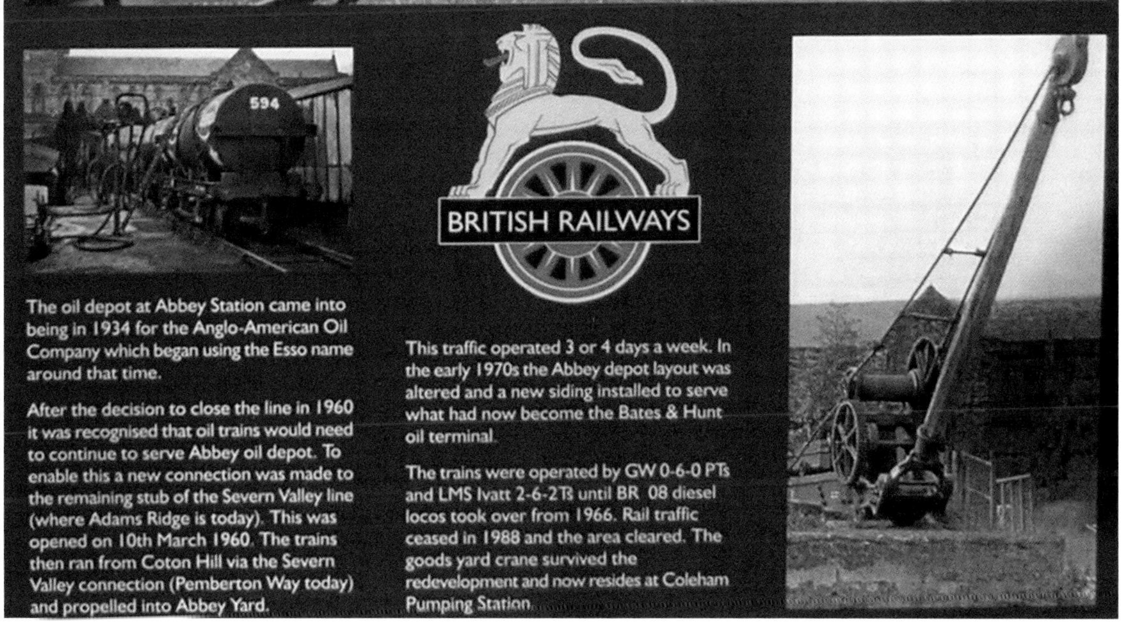

The oil depot at Abbey Station came into being in 1934 for the Anglo-American Oil Company which began using the Esso name around that time.

After the decision to close the line in 1960 it was recognised that oil trains would need to continue to serve Abbey oil depot. To enable this a new connection was made to the remaining stub of the Severn Valley line (where Adams Ridge is today). This was opened on 10th March 1960. The trains then ran from Coton Hill via the Severn Valley connection (Pemberton Way today) and propelled into Abbey Yard.

This traffic operated 3 or 4 days a week. In the early 1970s the Abbey depot layout was altered and a new siding installed to serve what had now become the Bates & Hunt oil terminal.

The trains were operated by GW 0-6-0 PTs and LMS Ivatt 2-6-2Ts until BR 08 diesel locos took over from 1966. Rail traffic ceased in 1988 and the area cleared. The goods yard crane survived the redevelopment and now resides at Coleham Pumping Station.

A coal train from South Wales, hauled by a Churchward 2-8-0, passes Coleham Locomotive depot in September 1959 on the approach to Shrewsbury, the wagons are all the antiquated wooden bodied variety, soon to be replaced by something more modern. A pannier tank is shunting at bottom left, and over in the depot is a Stanier 2-8-0. (Author)

Moving to the north end of the station on the same day, we see the 12.15pm all stations to Crewe hauled by a filthy dirty Prince Royal pacific No.46206, *Princess Marie Louise* and a Stanier Class 5MT 4-6-0. Pacifics were often used on filling turns between working the West Coast main line, and this no doubt explains its presence. (Author)

In complete contrast, on 10 September 1958 an immaculate No.46204 *Princess Louise*, having just been overhauled at Crewe, has arrived at Shrewsbury on an early morning stopping train. (Author)

It was on the curve on which we have just seen the Princess and the Stanier 5, that quite the worst accident in the history of the railways of Shrewsbury occurred. At around 2am on 15 October 1907, the North to West mail train attempted to take the curve at what was an estimated 70mph instead of the regulation 10mph. In the resultant crash, eighteen people died, four of them postal workers, three from Shrewsbury. The driver and fireman also died and no satisfactory reason for the accident was ever discovered. Amongst the wreckage is a GWR clerestory carriage. Buried beneath is the locomotive, No.2052 *Stephenson*. (Author's collection)

The King class did not make regular appearances at Shrewsbury until the late 1950s,. apart from a brief period around 1948/9, when No.6000 *King George V* was based at Bath Road depot, Bristol. In preservation days in November 1980 *King George V* has arrived from Didcot. The first vehicle is a Pullman, a most unusual sight at Shrewsbury, this being one which was owned by Bulmers of Hereford, a cider firm which also sponsored No.6000 for a while. (Author)

LNWR built *Prince of Wales* 4-6-0 No. 5605 'Pathfinder' about to depart Shrewsbury with a stopping train for Crewe, c1930. (LNWR Society)

2-6-2T No.5564 departs Shrewsbury for the Severn Valley line and passes the pioneer Castle, No,4073 *Caerphilly Castle* waiting to take over a South Wales express from a London Midland locomotive. Beyond in the station a Hall is about to depart with a Paddington express., 12 September 1959. (Author)

Shrewsbury 14 September 1961, No.6013 *King Henry VIII* is pulling into Shrewsbury passing the famous Severn Bridge signal box with the down Cambrian Coast Express. The class had long been associated with this train but had only recently begun to work beyond Wolverhampton. In the far left is the Manor which will take the train on into Mid-Wales. (Author)

Stanier 5MT 4-6-0 No.44679 stands on the through road at Shrewsbury, 14 September 1959, whilst Stanier 2-6-0 No.42980 pulls out with a Liverpool Lime Street express. Far right is a newly introduced DMU waiting to follow the Liverpool express, all stations to Crewe.
(Author)

0-8-0s of LNWR origin had been a familiar sight in and around Shrewsbury for some fifty years when this photograph was taken on 13 May 1956 of No.49206 and two others on Shrewsbury depot.
(Author)

SHREWSBURY • 93

No.1016 *County of Hants,* a Shrewsbury engine, has charge of a parcels train, which as was the nature of such trains, consisted of a great variety of vehicles, including a toplight full brake. On the adjoining platform is a recently introduced Swindon built Cross-Country DMU, which, along with the closure of a number of stations between Shrewsbury and Hereford, had brought about the cessation of steam hauled stopping trains on the line.
(Author)

The signalman at Sutton Bridge North Signal Box casts a watchful eye on No.70000 *Britannia* as it heads out of Shrewsbury for Hereford, in September 1981.
(Author)

One of the very few situations where Stanier 8F 2-8-0s were regularly rostered for passenger work was on the Central Wales line. No.48110 gets under way from Shrewsbury on 13 September 1958 with the 1.15pm to Llandovery. (Author)

No.7828 *Odney Manor* in charge of the Cambrian Coast Express, a turn with which it was associated for many years, curves towards Severn Valley Junction as it approaches Shrewsbury in 1964. (Author)

Shrewsbury shed, 13 September 1958. Far left is No, 1017 *County of Hereford*, not yet fitted with a double chimney, No.4037 *The South Wales Borderers* and No, 1013, *County of Dorset*, which has just received its double chimney. No.4037 was a most remarkable engine, for it was built as a Star in December 1910, became a Castle in June 1926 and was not withdrawn until September 1962. Whatever the GWR chose to officially call it, the reality is that it was a true rebuild. Renamed in the 1930s, it covered no less than 2.4 million miles in its long career, almost certainly a record for a steam locomotive in the UK. (Author)

Sentinel No.7109 *Joyce*, built at Shrewsbury in 1928. No.7019 was a pioneer and, as such was tried out by the LMS in 1927. It entered preservation in 1968 and eventually arrived at its present home, Midsomer Norton, where it is seen in August 2022. (Author)

One of the Sentinel company's famous S type steam lorries of the late 1920s. (Author)

Princess Coronation pacific 46252 *City of Leicester* is about to leave Shrewsbury with the 4.25pm stopping train to Crewe in the summer of 1958. The 37 mile run between Shrewsbury and Crewe was an ideal filling in turn for Stanier pacifics between the serious stuff of working the West Coast route, and thus one was just as likely to appear on an all stations stopper as a Black Five or a 2-6-4T. (Author)

Chapter 10
THE SHREWSBURY TO CREWE LINE

The LNWR was anxious to complete the link between Crewe and Shrewsbury as this would open up the prospects of establishing its own route to South Wales and the West of England, with the co-operation of the GWR.

The North of England to South Wales and West of England expresses which were such a distinguishing feature of the line, with carriages from a bewildering variety of starting points, as far north as Aberdeen and as far south west as Penzance, were instantly recognisable on account of their mixture of dark red and chocolate and cream painted carriages. During the night, through which these trains continued to run, there would also be teak-bodied LNER carriages and vans. Nationalisation in 1948 meant that red and cream became universal, but it was still possible to pick out the characteristics of one company's carriage against the others, regardless of its livery. Some ten or so years later BR Mark 1 carriages began to replace those of GWR and LMS origin. At the same time, diesel was taking over from steam so it all became rather less exciting. It has to be admitted that when the first generation of DMUs appeared on the 5.45pm out of Shrewsbury, the view through the driver's window opened up a never before available aspect of a familiar world.

Hadnall station, the first on the line to Crewe, opened, along with the line itself, in September 1858, and closed 102 years later. It regularly won prizes for its lovingly tended flower beds and pristine appearance. To alight after the long journey from South London and stand at the platform end as a South Wales express came hurtling non-stop through on the opposite track, just a matter of a few feet from where one stood, was a sensation both terrifying and thrilling.

Yorton always seemed much less busy than Hadnall. The yard consisted of just two short sidings, with often just one open coal wagon, so you could hardly call it a yard, unlike Hadnall which boasted several sidings and was always busy, and where there also seemed to be more passengers boarding or alighting from trains. Yet Yorton is still with us: it was school traffic which saved it. Yorton itself isn't even a village, just a pub and a collection of cottages, the nearest proper village being Clive, some ten to fifteen minutes' walk up the hill, beyond which is Grinshill with its quarry, the stone from which, apparently had a tendency to crumble into dust which got into carriages and thence passengers clothes, which upset first class passengers. No-one ever seems to have recorded the opinion of the lower orders.

Stopping trains in steam days were made up of both corridor and non-corridor carriages. Ex LNWR passenger locomotives had vanished by the time I could afford to buy Ian Allan ABCs in the late 1940s but I might well have seen and even travelled behind one as Prince of Wales 4-6-0s were actually still regularly employed throughout the war years on passenger trains between Shrewsbury and Crewe. However, my infant gaze would have barely bothered with these grubby, neglected, old fashioned looking inside cylinder veterans. Streamlined Princess Coronations were utterly different.

I recall two sightings of these unique looking mammoths. One was at Wem, painted black and at the head of a Crewe-bound stopping train. The other was from the bridge by Hardwick farm. Despite being filthy dirty, there was still enough of its pre-war striped blue livery visible for me to rush back to Aunt Agnes and ask her why the Coronation Scot was heading for Shrewsbury. Sadly, she was too busy feeding the hens to even consider the question, and Cousin Eric was away in the Navy.

Freight was different, the faithful old LNWR 0-8-0s seemed destined to go on for ever, and although they were finally replaced by Stanier 2-8-0s, the diesel revolution ensured the latter's reign was short lived. LNWR built carriages lasted a while longer, a three coach rake of toplight corridors, c1920, languished in the sidings at Hadnall for a while on their way to the breakers around 1956 and I once found myself hauled by a newly overhauled Prince Royal pacific in 1959, its newly applied Midland red paint barely dry, in an ancient cove roofed LNWR non-corridor carriage some 60 years old; goodness knows how it had managed to last for so long.

The proximity of Crewe, some nonstop 35 minutes away, meant that Hadnall, with its loop off the main line beside the signal box, was ideally placed for newly overhauled locomotives to ease themselves back into service and the lunchtime sight of three of them coupled together simmering away before making the return journey, was memorable. Stanier class 5s predominated, hardly surprisingly, whilst the really big express engines, Royal Scots and Stanier pacifics, either worked down singly or else were put in charge of an early morning stopping train to Shrewsbury, where they turned on the Abbey Foregate triangle and then headed back. In that context, I saw the ill-fated one time Turbomotive, No.46202, rebuilt as *Princess Anne*, come hurtling through Yorton with a West of England express, a few weeks before it met its end in the horrific Harrow and Wealdstone disaster.

The most important intermediate station on the line was Whitchurch. Certain north to west expresses stopped there, and it was the junction for the Cambrian line from Central Wales, and the LMR branch to Chester. I travelled on the latter, never the former, although I did watch an ancient Cambrian Railway 0-6-0 arrive from Wales one June evening in 1954 with three corridor carriages, including one almost new bow ended Hawksworth – the Western never seemed to mind so delegating its newest and finest to such duties.

I did make a journey on the Whitchurch to Chester branch. Motive power was a Midland Compound and 4-4-0s made something of a last stand here, Midland Railway type 2 class 4-4-0s also featuring into the 1950s, having replaced some of the last LNWR Precursors and George Vs. Although the date of that journey was 8 August 1953, the handwritten ticket bears the initials of the LMS, a company which had expired more than five and a half years earlier. This was, in terms of mileage, the shortest of the three possible ways to get from Hadnall to Chester, the second meant going to Shrewsbury and then along the Western main line, and the third taking a stopping train to Crewe and then aboard a North Wales express. Which was the quickest all depended on waiting times for connections. There were five stations between Whitchurch and Chester, serving scattered villages, including Tattenhall on the main line. The line closed to passengers in 1957 and to goods in 1963; for a while afterwards condemned wagons were parked on it prior to breaking up.

The Whitchurch to Oswestry line was considered the Cambrian's main line and was intended to be an important link with the West Wales coast, and, with the support of the LNWR by way of Whitchurch and Crewe, one that connected it with the rest of the UK railway system: certain trains continued on from Whitchurch to Crewe, complete with GWR motive power. The line was in direct competition with the GWR route from Chester by way of Ruabon, Welshpool to Barmouth Junction and the coast. Both routes carried a fair amount of holiday traffic in the summer peak season, but single track sufficed for most of the network. Ellesmere, an attractive town celebrated for its lakes, was the most important place between Whitchurch and Oswestry. The line closed in January 1965, leaving the line from Welshpool by way of Buttington, the onetime junction of the Whitchurch and Shrewsbury lines and owned jointly by the GWR and the LNWR, as virtually the only connection with the former Cambrian system and the outside world. Ellesmere station, in commercial use, is a rather remarkable survivor, complete with the main down platform and awning.

Whitchurch station, on the other hand, although it is still open to passengers, has been much reduced, to simple shelters on its two platforms, and virtually all evidence of its onetime status as an important junction has vanished.

One of the most celebrated of the documentaries made by the ground breaking GPO film unit in the 1930s was *Night Mail*, by Harry Watt and Ben Wright with music by Benjamin Britten and, perhaps, best known of all, the poem itself, *Night Mail* by W.H. Auden with which the film ends as the train descends towards Glasgow. One of the key scenes takes place at Crewe, where the London train meets that from North Wales. Mail bags cover the platform in this photograph at Crewe of a train hauled by 33026 *Sultan*, which has just arrived from South Wales by way of Shrewsbury, c1995. (Author)

Crewe was the railway town par excellence. One of the most famous locomotives built there is No.70000 *Britannia*, the very first British Railways standard steam locomotive. *Britannia* is seen at Crewe, after restoration to steam in 1991. (Author)

Stanier 8F 2-8-0 No.48110 passing Yorton on its way with a very mixed freight to Shrewsbury, and beyond, 13 September 1961. The first 8F was completed in 1938 and there were eventually 666 of them. (Author)

No.70017 *Arrow* at Hadnall in 1954, running in from overhaul at Crewe. Initially all the Britannias were overhauled at Crewe. It is coupled to a pair of Stanier 5MT 4-6-0s. After checking over to ensure all is well, they will return to Crewe and then to their home depots. (Author)

Above: Jubilee No.45629 *Straits Settlements*, at the head of the 11am Paignton to Manchester London Road, on 5 September 1958 is about to pass under the bridge carrying the lane from Hadnall village centre. (Author)

Opposite above: Three days earlier and a couple of miles in the Crewe direction, Patriot, No.45507 *Royal Tank Corps*, approaching Yorton, has charge of a returning Manchester to West of England train. There is the usual LMS/GWR mix of carriage; note how much wider the leading Hawksworth bow ender is than the Patriot's flat sided, coal piled high, tender. The Patriots were an interesting combination of LNWR, Lancashire and Yorkshire, and Midland Railway practice and, although some were Swindonised by Stanier, most lived out their time in original condition, still usefully employed on top link duties, until dieselisation. None was preserved but a replica is under construction. (Author)

Yorton in 1970. The two sidings have gone and there is a DMU stop sign. But otherwise, little has changed since LNWR days, the standard LNWR signal box, remains, extra high signal pokes above distant trees on the Bibby estate, and the board crossing is in place. Queen Mary, wife of George V used the station on her visits to her brother at Sansaw Hall, and it is said that there was a clause that the station would never be closed. (Author)

Take a close look at this view of Hadnall station, sometime in the late 1930s. Notice how neat and tidy it is. The staff took great pride in 'their' station and regularly won prizes for it. Which didn't stop British Rail closing it after an existence of 102 years in 1960. The yard is full of wagons, many of them private owner coal ones, Lilleshall, for instance, was one of the biggest Shropshire collieries. (Author's collection)

Diesels began to make their presence felt at the end of the 1950s and here an almost brand new Gloucester two coach unit is approaching Haston bridge, accelerating away from Hadnall station with the 5.45pm Shrewsbury to Crewe all stations., 4 September 1958. (Author)

Swansea-bound class 158 passing Hardwick farm between Yorton and Hadnall on its way from Manchester, 1990. (Author)

Yorton station, May 2022. The station building is now a private property, note the solar panels. 150283 is passing through, there being no passengers wishing to board or alight. (Author)

In the last days of steam, we see a scruffy looking Stanier 2-8-0, No.48332, its shed allocation, 9F (Heaton Mersey) roughly applied, at Crewe South depot in 1967. No.48332 lasted into the final year of steam, 1968, being sent to the scrapyard in February 1968. (Author)

A semi-fast service between South Wales and Manchester replaced the through West to North expresses which were diverted by way of Birmingham New Street. These were locomotive hauled by class 24s and 25s and here, in November 1983 a Manchester bound train is approaching Wem. (Author)

Looking south from Whitchurch c1978. Whitchurch was once an important junction, one branch, an LNWR line, heading north-westwards to Chester, the other, the Cambrian, coming in from Oswestry and Mid Wales, being that company's principal connection with the rest of the railway system, certain trains carrying on to Crewe. The Cambrian line to Oswestry, closed in 1965, branched off to the right. The line on the left led to the small engine shed. LMS-style semaphore signals remain, although not for much longer. A DMU is approaching from the Shrewsbury direction in the far distance. (Author)

Crewe in 1948 with representatives of the GWR and the LMS big passenger tank classes, companionably coupled up. GWR No.4154 would have worked in from Wellington by way of the Market Drayton line, whilst Stanier 2-6-4T No.42544, which has just acquired its British Railways number, might have been employed on any number of duties in the Crewe area. (Author's collection)

No.9017 stands in Oswestry station at the head of a stopping train for Machynlleth in the summer of 1954. Opposite cattle wagons are being shunted, for it is market day, whilst the impressive building on the left towering over the proceedings, is the onetime headquarters of the Cambrian Railway. (Author)

An attractive small town, Ellesmere is known as the capital of the Shropshire Lake district. It also attracts canal enthusiasts, for Thomas Telford, who built the Ellesmere, later Llangollen, Canal, had his office in the building seen here. (Author)

Two 45xx 2-6-2Ts head the Cambrian Coast Express along the coast north of Barmouth in 1947. (Phillips)

9017 was the last active 9000 class 4-4-0 and after various appeals it was secured in 1966 by Tom Gomm and brought to the Bluebell Railway, where it has been ever since. In 2009, No.9017 paid a visit to its old haunts to much acclaim and is seen here at Llangollen. (Author)

Llanfair and Caereinon Railway Beyer Peacock tank engine No.822 stands at Llanfair station in 1998. This 2ft 6in gauge line opened in 1903, was worked by the Cambrian Railway, absorbed by the GWR in 1923, who closed it for passengers in 1931 and freight in 1956. Part of the line was re-opened in August 1965 and there are plans to extend it to the main line station. (Author)

The Bibby line ship, *Shropshire*. Built on the Clyde, by Fairfield of Govan, in April 1959, it was sold to Asian owners in 1972. (Author's collection)

Wem, a typical LNWR design signal box and beautifully cared for until the very end which came soon after this photograph was taken in 1987. A Shrewsbury-bound class 153 is approaching. (Author)

Running in after overhaul at Crewe Stanier pacific No.46224 *Princess Alexandria* is about to make its final stop at Hadnall before terminating at Shrewsbury, 22 May 1959. *Princess Alexandria* was a Polmadie resident and would shortly be returning home across the border. (Author's collection)

Admirers talk to the crew of preserved No.46233 *Duchess of Sutherland* at Crewe in the spring of 2014 before setting off for Shrewsbury, Banbury and Paddington. (Author)

Chapter 11
Towards Chester

Chester is even nearer to Wales than Shrewsbury; and like it is a junction for railways to and from the Principality. Paddington expresses stopped at Gobowen, Chirk, Ruabon and Wrexham and, on arriving at Chester, reversed for the final stage to Birkenhead and for this relatively short run of 15¼ miles tank engines were generally considered adequate. Post 1945 that meant GWR design 51xx 2-6-2Ts and LMS 2-6-4Ts, both well proven machines with a potential for speed. The journey between Shrewsbury and Chester took around 75 minutes. The few through stopping trains, and not all of them stopped at every one of the 17 intermediate stations, took not all that much longer, around 90 minutes. Apart from the London trains, variety was provided by the daily through train from Bournemouth, and also one from the Kent and Sussex coasts, their green carriages ensuring that the liveries of all post-Grouping big four companies were regularly seen at Shrewsbury General's platforms. Chirk was where trains left England and on approaching the area around Wrexham a passenger who had slept most of the journey, were there such a being, might on waking have momentarily thought he had boarded the wrong train at Paddington and was in South Wales for this was coal country. There were a number of mines, some large, some small, which provided work for engines based at Croes Newydd GWR and Rhosddu Great Central sheds.

The story of the railway is inextricably linked with that of coal and none more so than in the Wrexham area. Gresford was the next station out of Wrexham on the Chester line, and it was here around 2am on 22 September 1934 that one of the most appalling disasters in our mining history occurred. An explosion, the cause of which was never fully explained, ripped through this deep mine, killing 266 miners. For anyone who has visited a coal mine, even one no longer worked but as an historical site and staffed by volunteer ex-miners, one struggles to imagine how men and boys – and, indeed, women and girls far enough back – could submit themselves to such appalling dangers in order to earn a living. And yet, when a mine was threatened with closure the workers fought to the end to preserve it and especially the way of life, the society that went with it. It is in some ways comparable to the soldiers in the trenches in the First World War who time and again went over the top, quite possibly not under any great patriotic urge but in support of each other.

Not only did families at Gresford lose fathers, brothers, sons and grandfathers, but 1,100 miners found themselves without work. The mine eventually re-opened, six months later, although it proved impossible to recover bodies, and back they went. Gresford Colliery only finally closed in 1973. The youngest victim was 14 years old, the oldest 69. One who survived because he was a lamp boy working on the surface, was 14-year-old Albert Reynolds, who lived on to the age of 100 and died in 2020. His father, a survivor of the battle of the Somme, and who had been awarded the military medal, was one of the victims. Albert recalled that he could always tell his father amongst all the other black faced miners returning to the surface, by ' his big grin'. At its peak, there were 38 pits in the Wrexham area, employing 18,000 men. The last, Bersham, closed in 1986, by which time its coal was hauled away not by steam locomotives, but diesel ones.

The presence of the LNER so far west in the Chester and Wrexham areas is explained by the fact that the GCR could have just as easily, or, perhaps, uncomfortably, been absorbed into the London Midland and Scottish Railway rather than the LNER in 1923. The LNER had its own station at Chester, Northgate, and although the LMR took it over in British Railways days, the distinctive presence of elderly locomotives of GCR origin remained, with 4-4-2 and 0-6-2T tank engines still finding work well into the 1950s, along with the Robinson 2-8-0s.

Chester General is a station with presence. Its architect, Francis Thompson, provided an impressive, Italianate face to the city centre, – it is Grade 2* listed. It expanded over the years and was not always the easiest big station to negotiate. Here and there sections were not properly roofed in, and in steam days in particular on a sunny day where the sun shone there was plenty of light but where it didn't it could be plunged into deep gloom. The twenty-first century has seen much improvement, £10million being spent by 2007, but the work continued beyond that date with better lighting, extra platform access and roof replacement.

Right from the beginning – it was completed in 1848, with later extensions – both the LNWR and the GWR provided a direct service to and from London, although the former, by way of Crewe and the main West Coast line was always slightly the faster, by a quarter of an hour or so,.

No.46202 *Princess Anne* pulls out of the north end of Shrewsbury with a West to North express in August 1952. Built in 1935 as the experimental steam turbine operated member of the Princess Royal class, although totally unlike all the others, it was nicknamed the 'Turbomotive'. Eventually rebuilt as a conventional pacific in the summer of 1952, it was destroyed in the horrendous Harrow and Wealdstone disaster of October 1952 when over 100 people were killed. Although not officially withdrawn until 1954, its demise allowed the construction of the one and only BR 8P pacific, the now preserved *Duke of Gloucester* at Crewe. (Shoesmith)

For a brief period between April 2008 and January 2011, a service ran between Marylebone and Wrexham. A Wrexham bound train of mark 3 carriages, hauled by a class 67, is seen at Gobowen, sporting traditional running in board and semaphore signals. Sadly the impossibility of getting round restrictions meant that it was bound to fail, the last run being on 28 January 2011. (Author)

Oswestry station, 2005. The headquarters of the Cambrian Railway, this handsome building has managed to survive various vicissitudes and is now in the care of the local authority. Passenger traffic through Oswestry ended in 1966 but freight continued to Llanyblodwel until 1988. The track is still in place and Cambrian Heritage Railways, who in August 2014 operated the first steam train out of Oswestry since 1965, are working to extend further along the line. (Author)

TOWARDS CHESTER • 117

An auto train worked a frequent service between the main line at Gobowen and Oswestry and is seen here departing Oswestry on 15 September 1959 with 0-4-2T No.1432 in charge. A DMU took over for the last six months of the service but could not save it; one doubts if was any more or less comfortable, and no faster than, its predecessor. Note the full freight sidings. (Author)

9xxx class 4-4-0 No.9022 departing Oswestry for Whitchurch in 1947. Although officially dating from the 1930s No.9022 is simply an amalgam of two late nineteenth-century specimens; the frames from a Bulldog and the boiler and cab from a Duke. (Author's collection)

Croes Newydd had the distinction of being the very last former GWR depot to operate steam. In this 1965 view on the left is a withdrawn 2251 class 0-6-0 minus numberplate, and beyond a BR Standard 4-6-0, a Churchward 2-8-0 and a 57xx pannier tank. (Author)

King George V pauses at Ruabon in driving rain on its way to Chester with an enthusiast's special from Didcot, 1988. (Author)

A drawing of the residents of Croes Newydd one Saturday afternoon in 1965. (Author)

In steam days, Wrexham was the most westerly point in the UK where locomotives from the GWR and the LNER met regularly, for it was here that the former Great Central line came in from Bidston. This latter somehow survived the great post-war cull of so many branch lines. In 2008, DMU No.150237 has called at Wrexham General having just left Wrexham Central – the indicator has not been changed – whilst a class 67 has charge of the Wrexham and Shropshire Railways train which has just arrived from Marylebone and will shortly return whence it has come. (Author)

This lady was just boarded the Marylebone-bound train at Wrexham General having come up earlier that morning from Banbury to visit a friend. Sadly, there were insufficient people in Banbury with friends in Wrexham, or vice versa, to make this wonderful service viable. (Author)

Looking down from a Chester-bound train on the Pontcysyllte viaduct which marks the border between England and Wales. (Author)

The aftermath of the appalling Gresford Mine disaster. (Author's collection)

Chester depot, 1939. At the head of the line is No.6812 *Gresford Grange*, completed at Swindon two years earlier. Immediately behind are three pannier tanks, two of them elderly open cab former saddle tanks, whilst opposite is a Dean Goods 0-6-0. Chester GWR shed was at the north west end of the station, the LMS one at the other. (Author's collection)

A Prince of Wales 4-6-0 heads past the walls of Chester with a train for the North Wales coast, c1930. (Author)

Canals are a prominent feature of the Chester scene and of late much has been made of the attraction of the canal-side world. The Chester Canal, linking the River Dee at Chester with Nantwich, opened in 1779, later became part of the Shropshire Union. The Chester Canal Trust was set up in 1997. (Author)

The LNWR managed to keep the company's presence in the Chester area alive throughout the 1930s and into the war years, with such as this Prince of Wales 4-6-0. (LNWR Society)

Class 67 No.025 propels the Holyhead to Cardiff via Shrewsbury Premier Service of mark 4 carriages, complete with restaurant car out of Chester, 5 September 2022. (Author)

The electrified Merseyrail system was eventually and most logically extended to Chester in 1993 and resulted in trains running every 15 minutes. A class 507 EMU is arriving at Bebbington, the station for Port Sunlight, with a Liverpool to Chester train in February, 2023. The 507s date back to 1978-80 and have been replaced by the 777s, the first of which arrived in the UK for trials in 2020. (Author)

TOWARDS CHESTER • **125**

A remarkably travel stained 66777 waits in the shadows beside the walls of Chester for the road westwards, 5 September 2022. (Author)

Chester Northgate station in 1955 with former Great Central Railway 4-4-2T No.67449 between duties.

Chapter 12
DESTINATION

Birkenhead is not the prettiest of English towns. In a February 2022 edition, the *Liverpool Echo* carried the headline, 'Birkenhead from New York of Europe to Town of Deprivation'. While it has fallen, more than once, on hard times, it has character and a certain vitality. The 'New York of Europe' stemmed from the links with Merseyside and the New World, millions from all over Europe passing through on their way to a hoped-for better life across the Atlantic. For those drawn to docks, ports, railway marshalling yards and engine sheds, the industrial and maritime heritage of Birkenhead is there for all to appreciate . It has some distinguished sons and daughters in many fields and can boast the first municipal park in England, whilst at its centre is the elegant Hamilton Square. If you still want to get away, it has excellent transport links with the world beyond.

Those in their post graduate year at in the art department of Liverpool University were encouraged to spend Friday afternoons at Birkenhead's premier cultural venue, the Lady Lever Art Gallery. We would find ourselves immersed in a unique cornucopia of amazing priceless artefacts, ranging from Ming china, Charles II's bed, all sorts of sculptures, and a number of delightful nineteenth-century soft porn, highly detailed paintings of naked young ladies disporting themselves. It was a nice end to the week. The LMS used to advertise excursions to the gallery and the soap works which had provided the cash for Lord Lever to finance both his gallery and his wonderful employees' village, Port Sunlight.

The first railway in Birkenhead, from Chester, arrived in 1847. Before that, George Stephenson had in 1830 carried out an abortive survey of the route between Chester and Birkenhead. The original terminus in Birkenhead was at Monks Ferry but this proved inadequate, and was replaced by Woodside in March 1878. It had a certain grandeur, although its site was cramped, eased in between a half mile long tunnel and the River Mersey. Marcus Binney has described Woodside as 'a station of truly baronial proportions and being worthy of any London terminus'. It was the only architecturally listed large Merseyside station, even though the porters and others who were paid to toil in its neglected gloom, might have demurred, Woodside had a pretty impressive front entrance but this soon fell into disuse, for the back, or side, door, of no great distinction, proved much more convenient for the ferries.

There had been numerous ferries criss-crossing the Mersey for centuries. Today just one survives, (and that is really for the tourists), all the others replaced by three tunnels, two for road vehicles, one for trains. Although the image is of Liverpool as being the main passenger port on Merseyside – and indeed it was for the Cunard and Canadian Pacific lines, until well into the 1960s – passenger liners bound for India and the Far East could be found tied up at the quayside in Birkenhead as well as fast, refrigerated cargo ships serving Australia and New Zealand, some of which also carried a few passengers, although they had to sidle over to the Pier Head, Liverpool to pick up them up . The jumbo jet, of course, put an end to passenger traffic by ocean liner, although not cruise liners which still call at Liverpool, nor ferries across the Irish Sea, which have survived and sail from Birkenhead. Containerisation ended Birkenhead's days as a premier commercial port, but at least the water is still there.

Ships entering the Mersey were destined not just for Liverpool or Birkenhead; they were also on their way to or from a port, one which for a while was the third – some say fourth – busiest in the UK. This was Manchester. It was the opening of the Manchester Ship Canal in January 1894, which brought about this transformation and in fact as late as 1959 it was at its busiest, ships of up to 10,000 tons being its bread and butter. But the decline when it came was dramatic and by 1982 Manchester Docks closed, to be partly replaced by Media City

None of these docks could have functioned without a huge network of railway connections, sidings and warehouses. The LNWR, the GWR, the Cheshire Lines Committee, the GNR, the Midland and the GCR companies were all involved. Manchester and Liverpool docks also operated their own railway systems. Birkenhead didn't. Uniquely, for the UK, Liverpool also operated an overhead railway. New York and Chicago have kept theirs but Liverpool's, suffering neglect during the Second World War, was found to be rusting away, the number of dockers, the mainstay of its clientele, was also falling away and so, before a similar fate could befall the overhead railway, it closed on 30 December 1956. Known as the Dockers' Umbrella, it was indeed a wonder. It enabled one to travel the whole length in comfort and view the fabulous 7-mile panorama of the docks. I had the spookily extraordinary experience of looking down from it on to the huge, upturned bulk of capsized liner *Empress of Canada* in the Gladstone Dock, which had caught fire on 25 January 1953 and then filled with the fire services' water and rolled over.

Trams also served the docks, Liverpool Pier Head rivalling London's Embankment as the ultimate mecca for the tram enthusiast. The last ran in 1956 but they nearly came back in the 1990s, the city getting to the point of buying the track, but then courage, vision and the money supply failed. Not so in Birkenhead, for although its present-day system is not much longer than a mile and one strictly for the enthusiast and the tourist, it can boast restored working vehicles from the Liverpool, New Brighton and Birkenhead systems, and brand-new ones imported, like so many bargains from the Far East, from Hong Kong. .

Three GWR 41xx 2-6-2Ts including No.4121, at Birkenhead Woodside, c1948. At this time, the big tank engines had a virtual monopoly of both stopping and express passenger trains between Birkenhead and Chester. (Author's collection)

Over at Liverpool Lime Street in August 1955 Princess Coronation pacific No.46250 *City of Lichfield* will shortly depart for Euston, but not until the driver, pipe firmly clenched between his teeth, has climbed back into his cab after a final track level check. (Author)

Liverpool Pier Head in the early 1930s. Liverpool was still building trams at the beginning of the Second World War and there were serious thoughts of restarting production as it came to an end. (Author's collection)

Port Sunlight remains a wonderful, beautifully maintained example of just what superb accommodation can be provided by an enlightened employer. (Author)

Fairburn 2-6-4T No.42164 provides the background for a fashion photograph featuring Miss Susan Butler near Rock Ferry in May 1965 as it surmounts the climb out of Birkenhead Woodside with the 2.30 pm to Paddington. (Author)

Above left: Woodside station shortly before closure. A Metro-Cammell DMU departs for Chester, a Stanier 5MT is at the head of a parcels train, and a porter goes about his business, November 1967. (Author)

Above right: One of the old Mersey Railway 0-6-4Ts No.5 *Cecil Raikes*, is still with us, having had an afterlife of some 50 years in a Derbyshire colliery and eventually passing into the National Collection. However, putting it on display at the Pierhead Liverpool as seen here in June, 1998 whilst perhaps appropriate, was not a terribly good idea and it is now under cover and in the care of National Museums, Liverpool. (Author)

Below: Into the 1960s, the Mersey ferries still offered serious opposition to the road and rail tunnels when it came to commuting from Birkenhead, Wallasey and New Brighton to Liverpool. Businessmen would regularly take exercise around the decks. A Manchester-bound freighter is being escorted by three tugs towards the Ship Canal and its destination in what was then the fourth busiest port in the UK, despite being 35 miles from the sea. (Author)

DESTINATION • **131**

An animated scene at Chester General station, 2010. (Author)

A Mersey Railway of American style clerestory carriages at Birkenhead Park c1920. (Author's collection)

The 9.30am Bournemouth to. Birkenhead, hauled by Crab 2-6-0 No.42981, heads through Bromborough near the end of its journey in September 1954. The Western and Southern Regions took it in turn to supply the stock, in this instance all the carriages appear to be of post-war Hawksworth design, apart from the refurbished 1930s vintage restaurant car. (Blenkinsop)

Birkenhead depot was the final home of a large fleet of BR Standard 9F 2-10-0s which were principally employed on steel trains to the John Summers works. (Author)

The derelict Birkenhead depot (8H), after the last steam locomotives had departed on 5 November 1967. It remained open for diesels until November 1985, and was demolished in July 1987. (Author)

Chapter 13
BIRKENHEAD EXPLORED

Before I knew better, I assumed that the town at the northern end of one of Britain's principal main railway lines would be substantial and, quite possibly, grand. Search as I might, I could not find its expected commercial centre with big department stores and the like. It had a quite grand, if rather gloomy, station with a big arched roof, alongside it a busy bus station and, adjoining both, a ferry terminal. And that was the problem, for on the other side of the Mersey, just ten minutes distant, was Liverpool, one of the great cities of the world with a world heritage waterfront, an accolade it has just lost (2022) on account of commercial pressures – very Liverpool – numerous department stores, two cathedrals, the world's all time most famous pop group, universities, some of the very first steel framed skyscrapers, two top of the range 1st division/Premiership football clubs, so many suburban cinemas that one could always find one where a Hitchcock film, for example, was playing – well, need I say more?

The area in and around Birkenhead docks necessitated a complex system of tracks worked largely by the GWR and LNWR, but there were also the Great Central, the Mersey and the Wirral companies, the two latter providing electric passenger trains. Given the rivalry of the GWR and the LNWR, it is perhaps surprising that there took place on 20 November 1860 the initial meeting of the Birkenhead Railway Joint Committee, the joint partners being none other than the afore mentioned companies. And, as T.B. Maund notes in his *The Birkenhead Railway*, published by the RCTS in 2000, 'Despite their fundamental competitive objectives, the two companies got on reasonably well together,' which, given how ferocious nineteenth-century competition could be, was surprisingly civilised. It probably helped that, to quote Maund again, that there was, for many years, 'Philip Henry Muntz … a permanent arbitrator between the purchasing companies' who was only called upon 'when some matter was in dispute'.

Although trains from Euston worked through to Birkenhead, these only consisted of a couple of carriages detached at Crewe or Chester from an express bound elsewhere, it was the Paddington route which was the principal one linking London with Birkenhead. In 1967, there were six through trains between Paddington and Birkenhead, that is until March of that year when they ceased, electrification of the 178 miles Euston to Liverpool route having been completed the previous year enabling the journey to be completed in a little over two hours.. In November of that year, Woodside was closed, from then on a change became necessary at Rock Ferry for all passengers to or from Chester or beyond. In 1993, third rail electrification was extended from Rock Ferry to Chester and in 1994 to Ellesmere Port.

My very first visit to Birkenhead was in 1953 arriving by way of a GWR designed 51xx 2-6-2T hauled train of non-corridor carriages from Chester, which deposited me at Rock Ferry where I boarded an electric train which took me under the Mersey to Liverpool Central, a journey of just ten minutes. Thus my first encounter with the centre of Birkenhead was from a distance, first through the windows of my train as it stopped at Birkenhead Central, which was not a lot being almost subterranean immediately before we plunged into the tunnel, or from across the water on the Liverpool bank of the Mersey. The Mersey Railway had opened in 1886 and became the very first in the UK to convert to electricity in 1903. The original electric trains were still in use in the early 1950s, the railway maintaining its independence until being absorbed by British Railways in 1948.

Mersey Railway trains were a great contrast to those I knew in the south. The whole ambiance reeked of being 'up north' – which, of course it was – with adverts specialising in products manufactured in Lancashire, promoting events alien to the London suburban experience, filled with people all with accents sounding like they were auditioning for a spot on a popular radio programme such as *Blackpool Night*; even the smell was distinctive. There was in addition a very transatlantic feel about the trains which was not surprising for American money had financed the electrification in 1903, the bogies and electrical equipment had been manufactured in the USA and the wooden bodies, although built by G.F. Milnes of Birkenhead as it happened, and later Hadley (we are probably almost all surprised to learn that railway trains were once built in that Shropshire town), were of pure transatlantic appearance with their plethora of high, narrow windows and clerestory roofs.

In the 1930s, the LMS decided to electrify the busy line from West Kirby to Liverpool. This, and the line to New Brighton, had always carried day trippers, holidaymakers and commuters into Liverpool and so in 1938 some very modern looking EMUs began work. Built by Metropolitan-Cammell Carriage and Wagon Company, nineteen 3-car units were provided. Constructed of welded steel, with guard-controlled air-operated doors, of open layout and plenty of standing space, they used the third rail but beyond Birkenhead North they switched to the Mersey Railway four rail system in order to reach Liverpool Central terminus. Their appearance put all other EMUs, particularly those of the Southern Railway, to shame, although their glamorous looks flattered somewhat to deceive for their riding was distinctly lively. However, they were considered sufficiently successful for twenty-four almost identical new sets to be built in 1956, type 503, well into the British Railways era, to replace the old Mersey Railway units. They had long lives, the last being withdrawn in March 1985.

In January 1956 I was sent to RAF West Kirby and spent eight weeks learning to march, bringing my boots and badges to a high polish, how to shave in freezing cold water and be generally converted from a hapless civilian to a finely tuned fighting machine, one fit to take charge of an Imperial, which was not, as it happened, a twin-engined fighter, but a typewriter. When released at weekends, my comrades-in-arms and I would travel on electric trains from Meols, which was the nearest station to RAF West Kirby, to Birkenhead or Liverpool and thus got to be familiar with both the modern, 1938 vintage, streamlined, sliding door, LMS-built EMUs and also the former Mersey Railway veterans, which were allowed out on Sundays to the West Kirby line where stations were spread much further apart and the potential for speed was much greater. This experience they clearly enjoyed for, presumably aware that their time was nearly up, they would bounce and rattle to tremendous effect with much joy de vivre.

I enjoyed it, too. Vastly better than being roared out, in a very uncouth manner, by drill instructor Corporal Abbot and threatened with an extra two weeks – horror upon horror – of square bashing at RAF West Kirby if I didn't 'smarten up'. Soon after the end of National Service, RAF West Kirby was declared redundant, razed to the ground, and returned to agricultural use, all evidence of its existence totally wiped out; it couldn't have happened to a more suitable candidate.

I am the first to admit that my experience of National Service was not universal. Some enjoyed it, visited places they had never imagined possible, and learned skills which stood them in good stead for the rest of their lives. Mine was that basically it was on balance a waste of time, a waste of money, too, for even the lowliest aircraftsman second class after two years in uniform could have made several pages long lists of the utter pointlessness of much that the Air Ministry regarded as the only way to run things. National Service was coming to an end, and I was mostly underemployed.

An upside was that I got to make close friends, during training and once qualified, with other clerk typists; especially Harry (Spike) Butterworth of Milnrow, Lancashire, with whom I shared a sense of humour and who also explained exactly why the Tour de France was the greatest sporting event on the planet, and Joe Hyam from Golders Green who had a BA from Oxford and, despite being the education clerk refused on principle – although I never fathomed precisely which one – to take an RAF trade test and who had 'Aircraftsman Second Class Hyam BA' emblazoned on his door, being better qualified than the officer in charge, and Corporal Jim a fellow bus and train spotter.

Clerk typist was quite possibly the least glamorous occupation in all the armed forces. I was grateful to the RAF for twice posting me to RAF establishments situated on the Western Region where I got to photograph one of the last two operational Star class 4-6-0s, and being delivered, after a week of kitting out at Cardington, to West Kirby by way of the joint GW/LMS line from Hooton, which closed soon after, an experience I didn't much savour at the time but now feel glad to have achieved. I was also able to watch the last Dean era tank engines engaged in shunting around the docks in Birkenhead, introduce comrades to the delights of riding a Liverpool tram, also about to disappear, and sailing up and down the Mersey on the then frequent ferry service from New Brighton, Wallasey and. Birkenhead to the Pier Head, Liverpool. It wasn't all bad.

Birkenhead depot was actually two establishments, their sheds side by side and with a shared access to and from the main line. Around forty locomotives lived at the GWR shed, code BHD, and at the LMS/LNWR one, code from 1935 6C, from 1963 8H, which was home to as many as seventy at times. In British Railways days, in 1951, the LMR took control. GWR classes were still seen there, but in ever decreasing numbers and eventually disappeared by 1960, so that Paddington expresses were from then on usually worked by LM design 2-6-4Ts, although Crab 2-6-0s were not unknown. By the time I began to nose around the shed, never challenged, c1963, that most impressive of all the BR designs, the 9F 2-10-0s, was dominant, having moved along the coast from the former GCR Bidston depot where they monopolised the Bidston Dock and Shotton ore trains. There they had replaced the long-lived Robinson 2-8-0s. Compared to them, the 9Fs had very short lives, but this was through no fault of their own but rather the machinations of politicians and management who had decided that steam should be got rid of with indecent haste, unlike France and Germany, for example, where a more measured approach was pursued.

Hindsight is a wonderful thing, but the truth is that working on the railways was no longer seen as the desirable career it had once been, there were jobs on Merseyside – containerisation had not yet transformed the docks and wiped out most of the employment opportunities – which paid better than working in the filthy, neglected confines of an out of date steam locomotive shed. Thus, the 9Fs seldom saw a cleaner's rag, but they were still impressive machines, although there were a few oddities such as those converted from the Franco-Crosti abortive, double boilered experiment, which came far too late. That excellent railway and wildlife artist David Shepherd bought one of the Birkenhead 9Fs, No.92203, named it *Black Prince* and ensured its preservation.

Birkenhead shed closed to steam in November 1967, the last diesel quit in November 1985. Ian Allan organised a final steam special, the 'Zulu', between Paddington and Birkenhead on 4 March 1967, with No.7029 *Clun Castle* working between Didcot and Chester and a BR Standard Class 5 4-6-0 between Chester and Birkenhead.